An Early Start to Technology from Science

Roy Richards

SIMON & SCHUSTER

LONDON • SYDNEY • NEW YORK • TOKYO • SINGAPORE • TORONTO

Text © Roy Richards 1990
Design and artwork © Simon & Schuster 1990

First published in Great Britain in 1990 by
Simon & Schuster Ltd
Wolsey House, Wolsey Road
Hemel Hempstead HP2 4SS

Printed in Great Britain by
BPCC Paulton Books Ltd

British Library Cataloguing in Publication Data

Richards, Roy
    An early start to technology from science
    1. Physical sciences
    I. Title
    600
    ISBN 0–7501–0033–8

Series editor: John Day
Editor: Julia Cousins
Design and artwork: Anna Hancock
Photographs: LEGO UK Ltd

ACKNOWLEDGEMENTS
The author and publisher warmly thank Margaret
Fisher, Lee Church of England Primary School,
London SE13, for permission to reproduce on page
92 her choreography for the Cotton Mill Work
Dance, and for her help and advice on its accurate
presentation.

They also thank LEGO UK Ltd for its special
permission to use the LEGO® Dacta photographs
reproduced on page 91. LEGO® is a registered
trademark belonging to the LEGO Group.

The Carpenter Song on page 61 is reproduced by
kind permission of the British Broadcasting
Corporation.

The links between technology and science are close. Underlying scientific principles fuel the achievement of technological goals. For young children the understanding of such principles begins with the gathering of experiences, and for them the understanding of technological goals often begins with making something. Understanding and making have resulted in a world where we take for granted hopping on a plane to fly to exotic lands, watching an astronaut in space courtesy of our television sets, or just using a calculator to tot up our purchases in a supermarket. It is a technological world that young children grow up in.

There are various aspects of science that link closely with technology, and these are the subject of this book. The book presents a wealth of practical experience which will involve children in looking at the **structure** *of things, at the* **materials** *things are made of, at* **forces** *and at* **energy** *(usually in terms of what makes things go), and at how things are* **controlled.**

The book introduces children to the processes of:

- **exploration** *of their environment in order to gather experiences at first hand*
- **manipulation** *of objects and materials*
- **observation** *of things around them*
- **questioning** *and arguing about things*
- **testing** *things out, indulging in simple problem-solving activities*
- **looking** *for pattern and relationship*

Some situations demand the making of things, whether it be a strength testing machine to find out how strong fibres can be, or a device to carry a line across the river in the building of a suspension bridge. This will involve children in:

- **researching** *the problem*
- **considering** *possible solutions*
- **choosing** *the best solution*
- **designing** *a device or making a plan on paper*
- **making** *the device or carrying out the plan*
- **testing** *the device or carrying out the plan*
- **improving** *the device or the plan – or scrapping them and seeking a new solution*

The twin themes of gathering experiences to help bring about understanding, and designing and making are interwoven throughout the book. At the same time an attempt is made to fit the experiences into the primary curriculum so that subject barriers are crossed, and the work the children do is part of the integrated approach to learning common in primary education.

Safety in schools

All the activities in this book are safe provided they are properly organised and supervised in accordance with the recommendations of the DES, the Health and Safety Executive, the Association for Science Education, and local authority regulations. Any teachers who are uncertain about safety in scientific and technical work should consult their LEA advisers. They should also read *Be safe: some aspects of safety in science and technology in primary schools*, published by the Association for Science Education.

Red triangles

Some activities in this book do require extra care and attention. They are marked with a *red triangle*. Under no circumstances should children be allowed to pursue them unsupervised, particularly during breaks.

Always pack away potentially dangerous apparatus and chemicals immediately the activity is over.

The National Curriculum

The companion volumes, An Early Start to Science and An Early Start to Nature, cover many of the attainment targets of the science programme. The present book concentrates on three of the physical science attainment targets:

AT6	Types and Uses of Materials
AT10	Forces
AT13	Energy

A wealth of experiences are suggested which will help children develop an understanding of concepts in each of these areas. At the same time, the underlying principles of AT 1, Exploration of Science, permeate the activities.

A host of things are included in the present book which can be designed and made, and which can be fitted into the context of many topics commonly carried out in the design and technology field. There is a contribution to each of the five attainment targets of the Design and Technology Curriculum. There is also much that can be used under the four groupings of the programme of study for Design and Technology. Namely under

> *Developing and Using Systems*
> *Working with Materials*
> *Developing and Communicating Ideas*
> *Satisfying Human Need*

I have gathered together much that is tried and true, and added things that are new. I hope that it will benefit all primary teachers and will help to take things forward as we enter the new era opened up by the National Curriculum.

Roy Richards

Crossing rivers and gorges has always presented problems. Give the problem to children. Make a 'stream' from blue sugar paper, or PE mats, or two lines chalked a short distance apart.

Can you cross without getting wet?

Jumping

Can you jump across?

Take care!

Swinging on a rope

Can you swing across?

Using stepping stones

Improvise with carpet tiles, or table mats, or lino tiles, or pieces of card.

Using a plank

Using a PE bench

Name	Crossed	Fell in
Jo		
Karen		
Sien		
Andja		

Name	How I crossed the stream
Sally	
Anna	
Tom	
Winston	

Increasing the load

Get increasing numbers of children to stand on the 'bridge'. What happens?

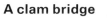

‡ sag

Measure the distance between the centre of the plank and the floor. This will give you the amount of sag.

‡ sag

‡ sag

Adding support

What happens with the central brick in place?

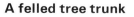

Number of children	Amount of sag
Ann	mm
Ann + Winston	mm
Ann + Winston + Scott	mm

A clam bridge

This is one of the earliest bridges. A single slab of stone stretching from bank to bank, making a single span.

A felled tree trunk

A clapper bridge

This consists of a number of slabs of stone, and a number of spans.

Different shapes

Cut some thin card into a strip measuring 650 mm × 150 mm and make a bridge. Use coins as a load. How strong is the bridge?

Try to strengthen the bridge by using pieces of card folded in different ways. Make sure you keep each piece 650 × 150 mm for a fair test. Make sure you keep the gap between the supports 450 mm.

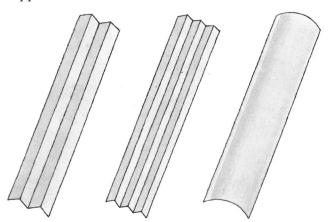

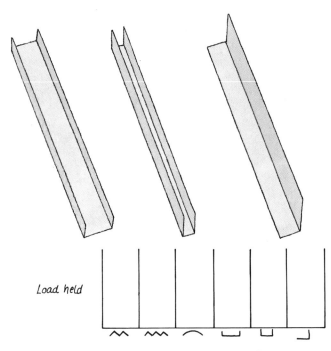

Load held

Different thicknesses, different materials

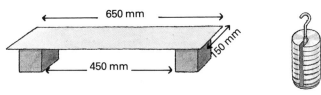

650 mm

450 mm

150 mm

Try:

newspaper   sugar paper   thin card   hardboard

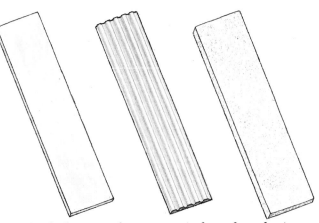

balsa wood   corrugated card   polystyrene

*You'll need a heavy set of weights.
Keep a record of the results.*

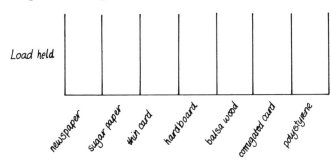

Load held

newspaper   sugar paper   thin card   hardboard   balsa wood   corrugated card   polystyrene

*Find the mass of each bridge.
Is there a correlation between mass and strength?
Is the bridge stronger when the beam is balanced on its edge?*

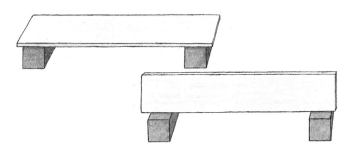

Different widths

Does the width make a difference to the strength?

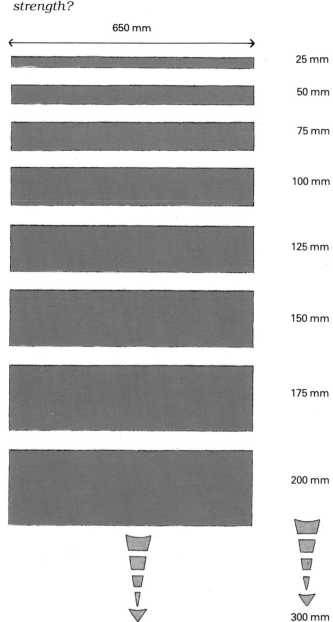

650 mm

25 mm

50 mm

75 mm

100 mm

125 mm

150 mm

175 mm

200 mm

300 mm

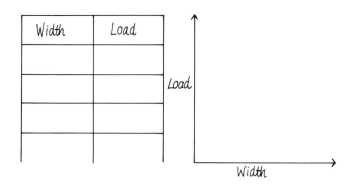

Width	Load

Triangular trusses

Beam bridges are often supported by trusses on each side.

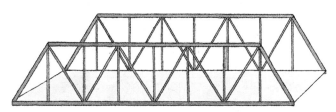

Investigate how trusses give strength.
Using a ruler, draw a series of equi-distant lines on a piece of card.

Cut out the strips. Make each 150 mm long and punch a hole in each end.

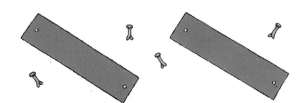

Join four strips into a square with paper fasteners.

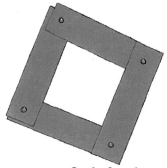

How many ways can you find of making a rigid shape?

Try

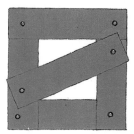

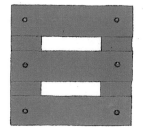

Which is best?

Make an arch bridge

Can the smallest child in the class step safely across? Will the bridge hold heavier people?

Take care!

The 'deck' of the bridge can be made from a sheet of hardboard 600 mm × 400 mm. But a strong tray or anything similar will do.

The 'arch' of the bridge is made from heavy-duty corrugated plastic sheet as shown below. If necessary, use two or three strips of plastic sheet, one laid over the other to make the arch much stronger.

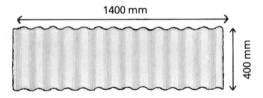

1400 mm

400 mm

A single-span packhorse bridge

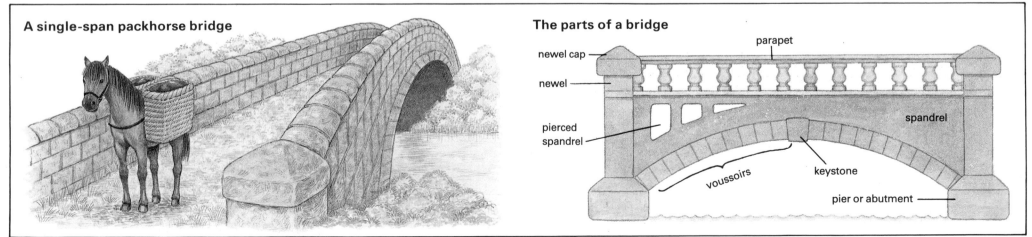

The parts of a bridge

parapet

newel cap

newel

spandrel

pierced spandrel

voussoirs

keystone

pier or abutment

Using this apparatus, investigate the effect of changing the height and span of the bridge on its strength.

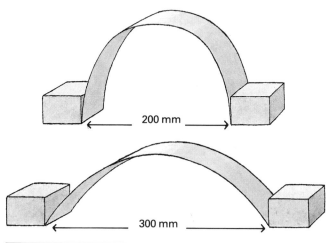

Different spans

How does the span alter the strength of the arch? Use a piece of card 450 × 100 mm.

200 mm

300 mm

Height	Span	Load
	200 mm	
	300 mm	

Different heights

What happens if you keep the span constant but vary the height? You will need different lengths of card for this.

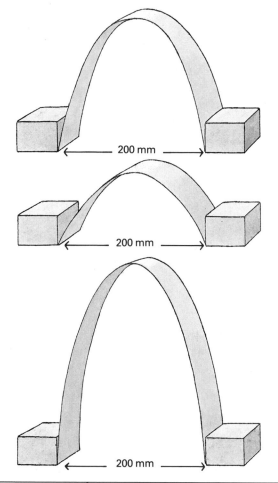

200 mm

200 mm

200 mm

Height	Span	Load
	200 mm	
	200 mm	

Beam and arch

Often the arch supports a beam. This type of bridge is common on motorways.

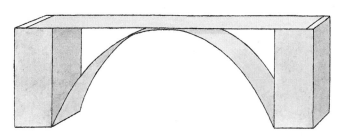

Does this make a stronger bridge?

Again, try varying the height and span.

Keystone arch

Draw on tracing paper the pattern of stones used in a keystone arch.

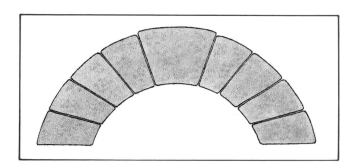

Use this to transfer the pattern to a block of polystyrene. Now cut out the 'stones' and construct the arch.

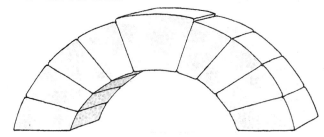

Cantilever bridges are common on motorways. They consist of a long concrete beam resting on brackets at each end, called cantilevers. Each cantilever is rather like a strong shelf bracket.

Make a cantilever bridge

Use balsa wood, and these templates, to make model cantilever bridges.

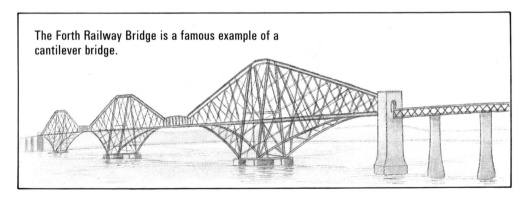

Take care! No broken bones. The child sitting on the 'beam' of the bridge must always get on last, and off first.

The Forth Railway Bridge is a famous example of a cantilever bridge.

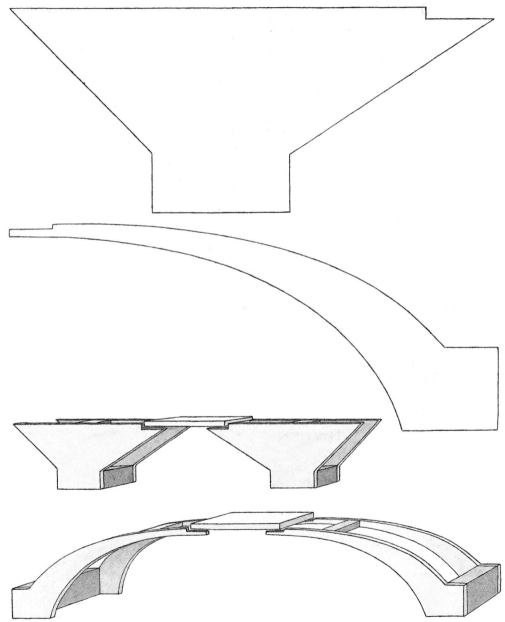

Where are the forces?

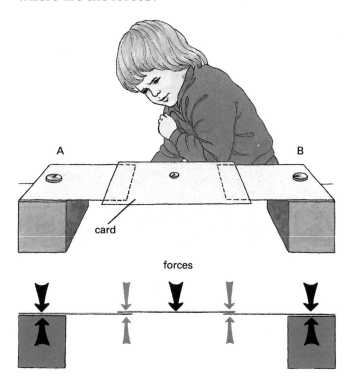

card

forces

Put a load at A and the same load at B. What is the greatest load you can put at the centre?

As you increase the central load, you will eventually need to increase the loads at A and B. How good can you get at predicting these?

Make a table of the results.

Central load	Load at A		Load at B	
	Predicted	Actual	Predicted	Actual

Vary the length of the central span

Keep the length of the sides constant.

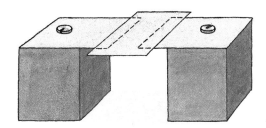

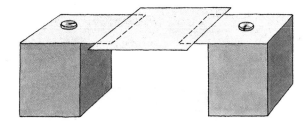

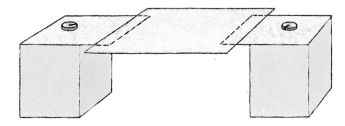

Length of central span	Load

Vary the length of the sides

Keep the length of the central span constant.

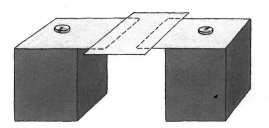

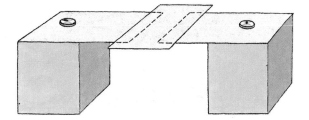

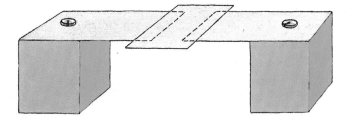

Length of side	Load

Build a suspension bridge

Work on a large scale.

tie main ropes here
to stop them
slipping

Use balsa wood towers to make a smaller model.

5 × 5 mm

500 mm

100 mm

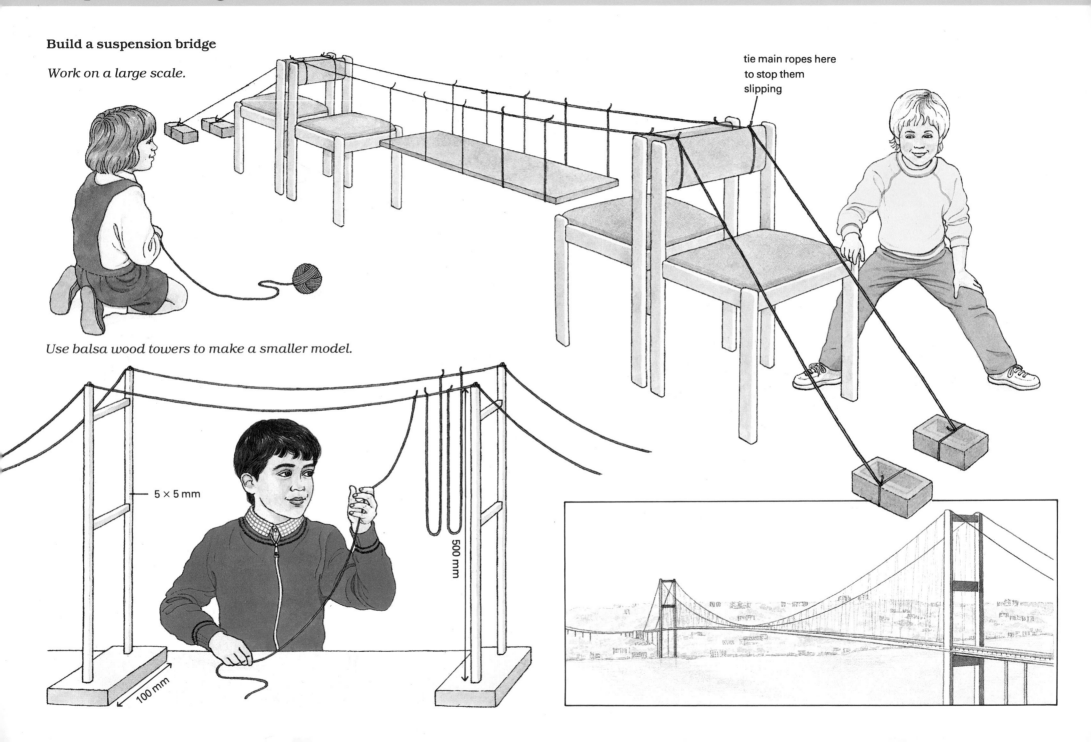

The drawbridge of a castle

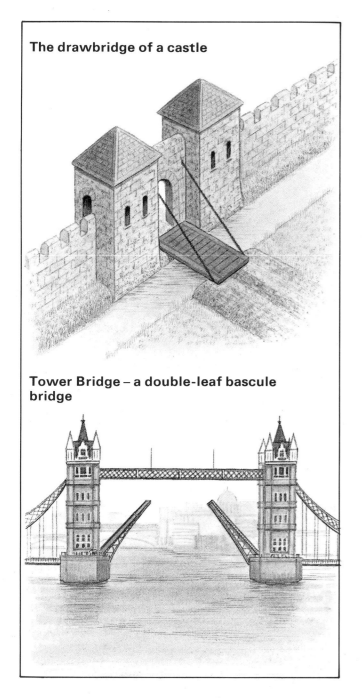

Tower Bridge – a double-leaf bascule bridge

Design some lifting drawbridges

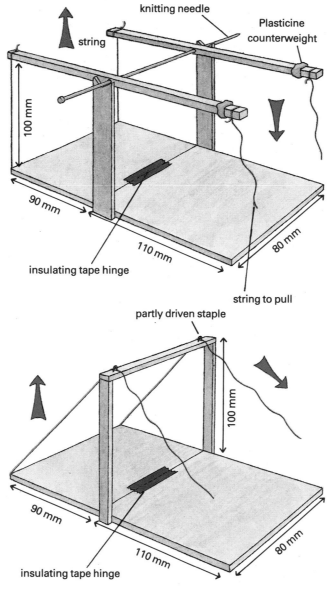

knitting needle

string

Plasticine counterweight

100 mm

90 mm

110 mm

80 mm

insulating tape hinge

string to pull

partly driven staple

100 mm

90 mm

110 mm

80 mm

insulating tape hinge

Design a hand operated single-leaf bascule bridge

These can be found along some canals.

closed

open

In building a suspension bridge, engineers need to get the first line across and then spin cable to and fro.

Take a line across by boat

You will need

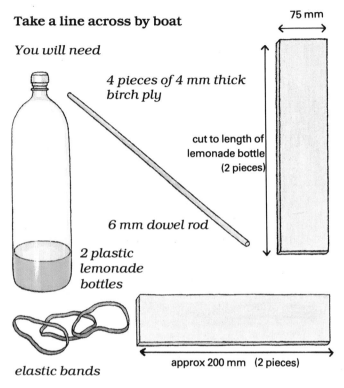

4 pieces of 4 mm thick birch ply

75 mm

cut to length of lemonade bottle (2 pieces)

6 mm dowel rod

2 plastic lemonade bottles

elastic bands

approx 200 mm   (2 pieces)

Drill a hole through the centre of the cap and the base large enough to take the 6 mm rod.

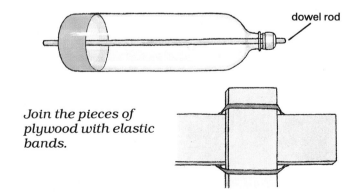

dowel rod

Join the pieces of plywood with elastic bands.

Using a sail

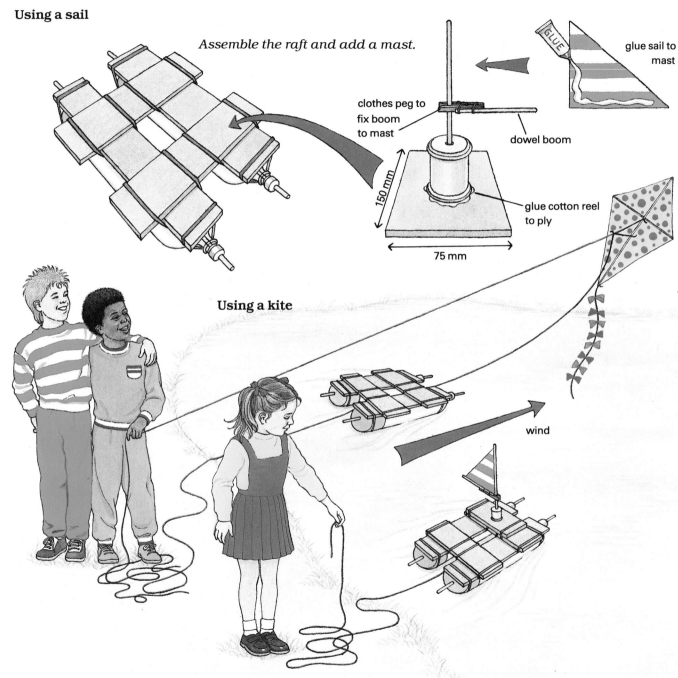

Assemble the raft and add a mast.

glue sail to mast

GLUE

clothes peg to fix boom to mast

dowel boom

150 mm

glue cotton reel to ply

75 mm

Using a kite

wind

Invent other ways of getting the first line across a 'river'.

By rocket

You will need

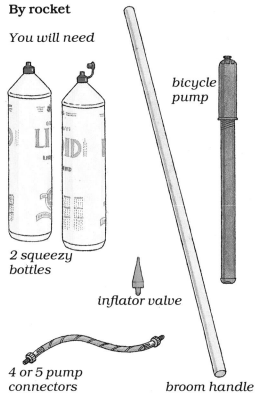

2 squeezy bottles

inflator valve

4 or 5 pump connectors

bicycle pump

broom handle

Cut the top from one of the squeezy bottles.

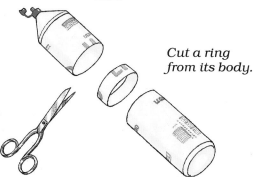

Cut a ring from its body.

Cut fins from the remaining part of the body. You will need three fins.

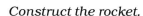

Construct the rocket.

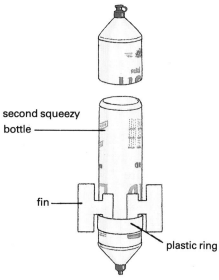

second squeezy bottle

fin

plastic ring

Make the launch site.

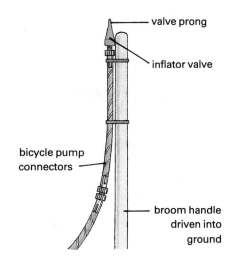

valve prong

inflator valve

bicycle pump connectors

broom handle driven into ground

Firing the rocket.

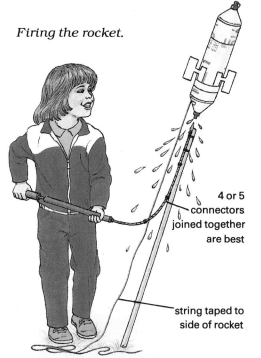

4 or 5 connectors joined together are best

string taped to side of rocket

Pour a cupful of water into the rocket. Place the rocket firmly on the valve prong. Pump in air until the rocket takes off. Wear your mac!

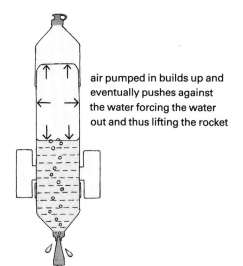

air pumped in builds up and eventually pushes against the water forcing the water out and thus lifting the rocket

By kite

By arrow

Take care!

notch

ash twig

notch

garden flower stake

Plasticine

Who can build the tallest tower?

Use different materials

shoe boxes

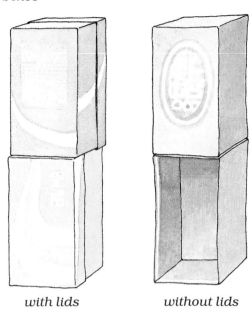

with lids   *without lids*

Does propping up the base help?

Use different objects as the base

What happens when the base is smaller in cross-section than the bricks?

What effect does the height of the base have?

Who can build the tallest leaning tower?

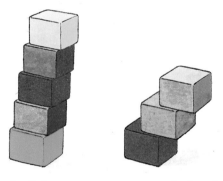

Who can build a tower most out of plumb?

Children will be experimenting with the centre of gravity, with the relationships determining the distribution of forces in their structures, as well as exploring space in a vertical direction as they build.

LEGO

yoghurt pots

paper cups

matchboxes

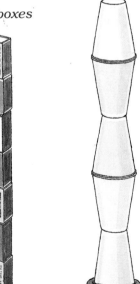

With newspaper tubes

Rolled newspapers make strong struts. Roll each newspaper tightly and bind with sticky tape.

Use these to build a tower.

How tall a tower can you build?

Bind the joints with string or sticky tape as you go.

With straws

Given 25 drinking straws, 50 pins, a softwood base and a marble, who can make the tallest tower to support the marble?

Using the same materials, who can make the longest structure built horizontally?

Will this hold a marble?

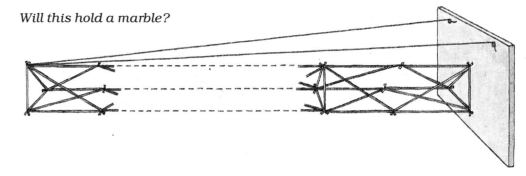

Cubes

A cube is a common, basic and simple structure.

How many nets can children devise to construct a cube?

The Platonic solids

The Greeks used a cube to represent the earth, and the other regular solids, tetrahedron, octahedron, dodecahedron and icosahedron, to represent fire, air, the universe and water, respectively. Use plastic templates to construct these nets.

Tetrahedron (fire)

net of a tetrahedron

flap
fold   fold   fold
flap

Octahedron (air)

net of an octahedron

flap
flap
flap
flap

Icosahedron (water)

net of an icosahedron

flap   flap   flap   flap   flap

flap

flap   flap   flap   flap   flap

Dodecahedron (universe)

rhombicuboctahedron

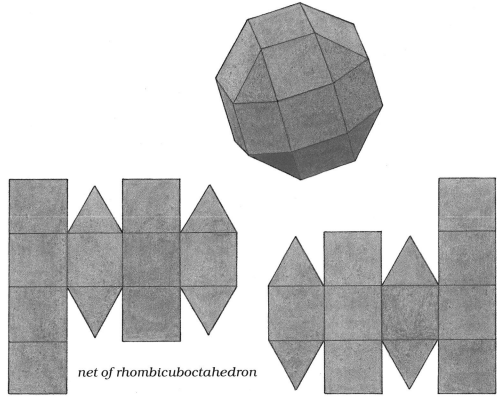

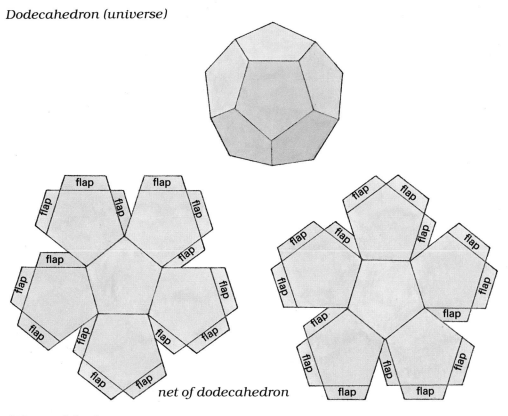

net of dodecahedron

net of rhombicuboctahedron

Other polyhedra

There are lots of other interesting polyhedra.

stellated octahedron

truncated octahedron

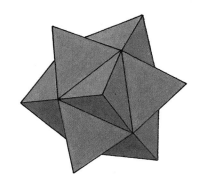

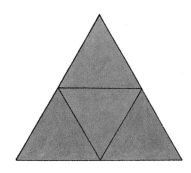

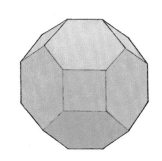

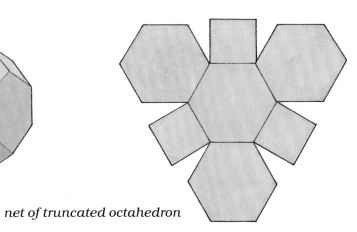

This is constructed from a series of trihedral pyramids.

net of truncated octahedron

The paper engineering found in popular pop-up greetings cards gives children scope for creative invention.

Below are some ideas to try out.

Fireplace

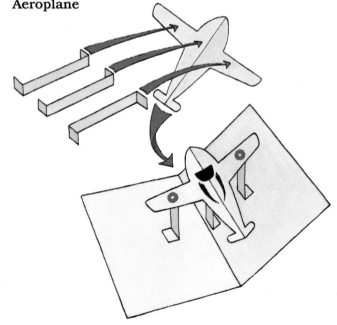

Aeroplane

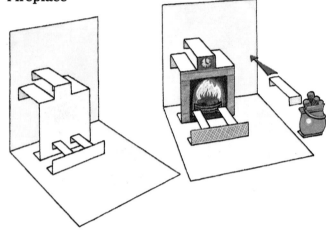

Baby in a playpen

Balloon

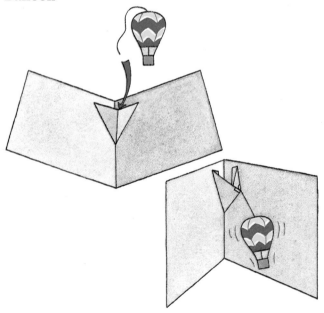

Castle

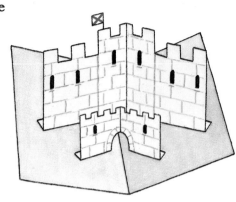

Rainbow

Butterfly

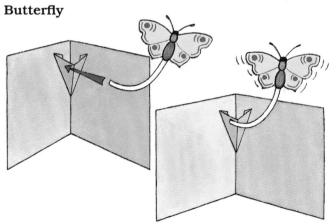

Cat and mouse

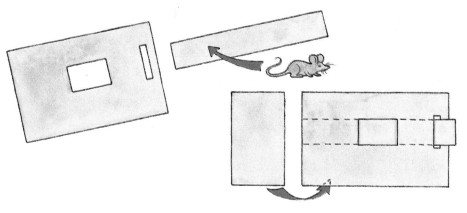

Snowstorm

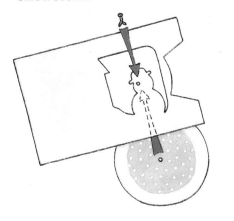

Flickering fire

Ringing bell

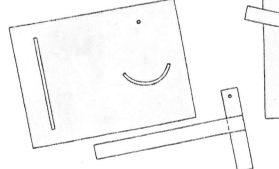

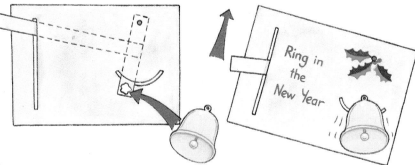

Father Christmas arrives

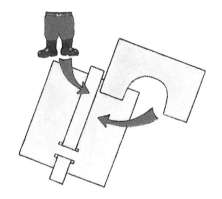

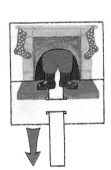

The structure and strength of plant stems are worth investigation. Make a collection from garden herbaceous stems and weak climbing stems to the woody stems found on shrubs and trees. Order them from weakest to strongest.

Look at plants in cross-section too.

Plants in cross-section

Cylindrical

Common in many plants from dandelion stalks to tree twigs.

Angular

For example, a celery stalk.

Square

Common in the deadnettle family.

Triangular

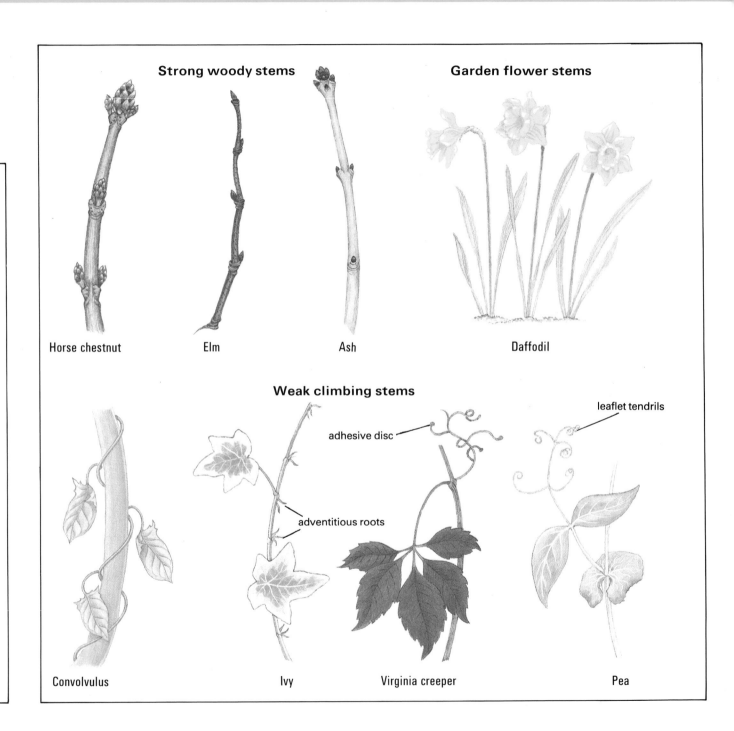

Strong woody stems

Horse chestnut   Elm   Ash

Garden flower stems

Daffodil

Weak climbing stems

leaflet tendrils

adhesive disc

adventitious roots

Convolvulus   Ivy   Virginia creeper   Pea

Using stems to print patterns

powder paints

Use the cut ends of different types of stem to print patterns.

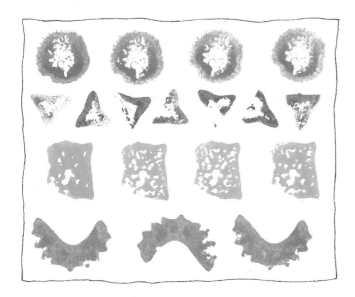

Examining stem strength

Investigate whether different shaped cross-sections affect stem strength. Use cartridge or sugar paper to make a 'stem' and add progressively larger loads until the 'stem' crumples.

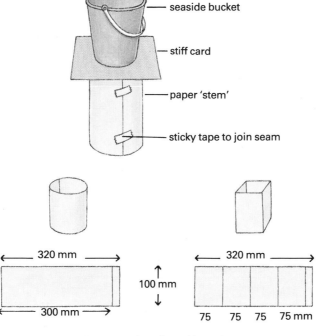

- seaside bucket
- stiff card
- paper 'stem'
- sticky tape to join seam

320 mm
300 mm
100 mm

320 mm
75 75 75 75 mm
100 mm

20 mm overlap allowed in each case

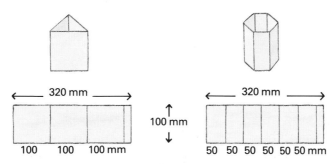

320 mm
100 100 100 mm
100 mm

320 mm
50 50 50 50 50 50 mm
100 mm

Which shape is the strongest?

Stem shape	Load
○	
□	
△	
⬡	

What happens with thicker 'stems'?

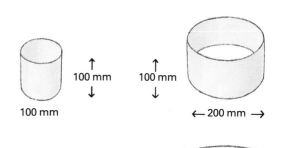

100 mm
100 mm
100 mm

← 200 mm →

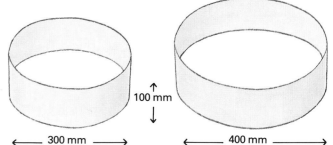

100 mm

← 300 mm →
← 400 mm →

Stem thickness	Load
100 mm	
200 mm	
300 mm	

The need for a skeleton

Our skeleton holds us up. The skull, rib-cage and pelvic-girdle also afford our internal organs a certain amount of protection.

During PE let the children pretend they do not have a skeleton and are floppy, rather like a rag doll.

Discuss how we need bones to hold us up.

Now let the children pretend that they have a skeleton, but that it is jointed only at the shoulders and hips.

Now try walking normally, and try some standing jumps.

Discuss the need for and use of joints.

Plot the joints in the body

Draw round a child to get a body outline.

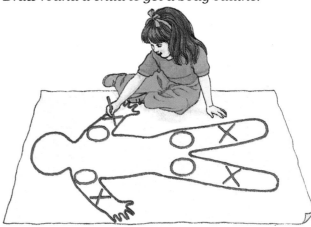

Mark in the ball and socket joints with a circle, and the hinge joints with a cross.

Collect x-ray pictures

Collect bones with different joints

These may be found in poultry and from the Sunday joint.

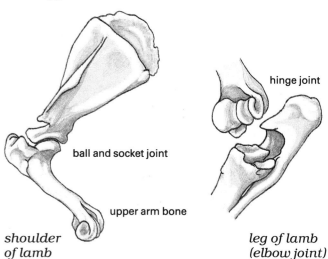

hinge joint

ball and socket joint

upper arm bone

shoulder of lamb

leg of lamb (elbow joint)

Make a model skeleton

Use a broom handle as a mould to roll papier-mâché tubes from newspapers and wallpaper paste. Use card to make the head, chest and pelvic girdle.

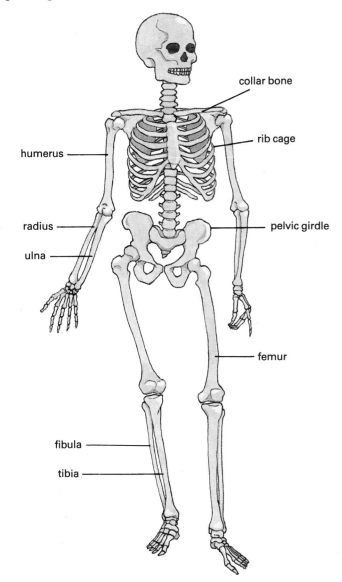

collar bone

rib cage

humerus

radius

ulna

pelvic girdle

femur

fibula

tibia

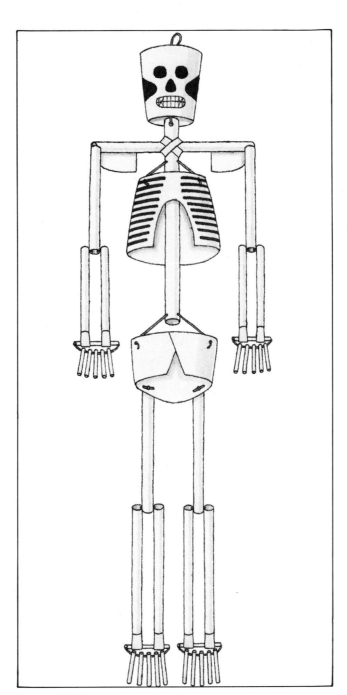

Slipped disc

Mimic the effect of the cartilage discs in the backbone.

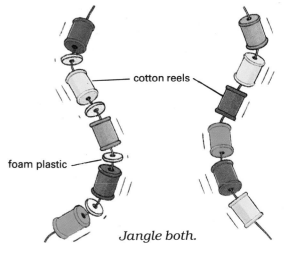

cotton reels

foam plastic

Jangle both.

Examine articulation.

Thread a drinking straw on string.

Cut the straw into three pieces. Rethread it.

Cut into smaller pieces.

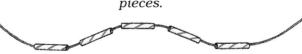

Relate this to a highly jointed backbone.

Keeping our balance

We are constantly, if only subconsciously, adjusting our balance. Our muscles work hard overcoming gravity. Let children explore balance in themselves.

Balance on a beam.

Now balance holding two plastic bottles full of sand.

Now try it with a broom handle or window pole.

When was it easiest?

How well do four-legged creatures balance?

Make some four-legged 'animals' from cardboard boxes, toilet and kitchen roll tubes.

Which is easiest to knock over, the tall beast or the short beast?

Vary the position of the legs.

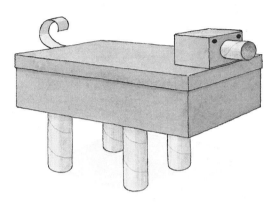

Which positions of the legs give greatest stability?

Try it with two-legged creatures.

Make some balancing toys

cardboard figure

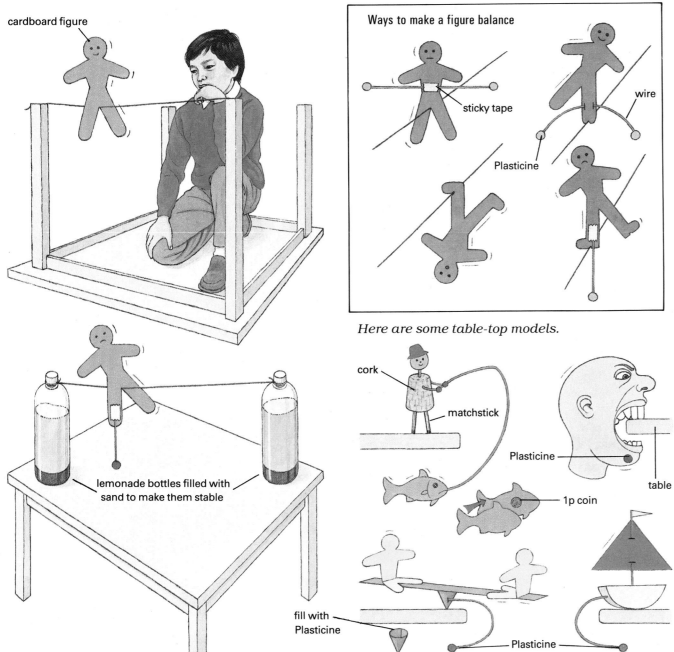

lemonade bottles filled with sand to make them stable

Ways to make a figure balance

sticky tape

wire

Plasticine

Here are some table-top models.

cork

matchstick

Plasticine

table

1p coin

fill with Plasticine

Plasticine

And here is another type of 'balancer'.

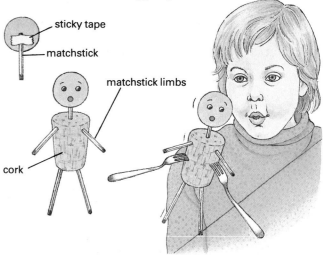

sticky tape

matchstick

matchstick limbs

cork

Blow gently from behind to make it move.

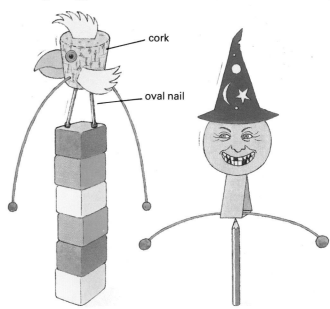

cork

oval nail

Investigate the effect of changing the size of the Plasticine bobs or varying the length of the wire. Discuss how a low centre of gravity gives stability and balance.

How many ways can you find to 'carry' someone across the school hall or playground? Which are the best methods?

Children should begin to get a feel for overcoming friction.

Guide and pull gently.

Take care!

Push backwards

invert

small tin lid

marbles proud of the rim

large tin lid

Discuss how lifting, dragging, rolling, wheels, and so on, help movement.

On playground slide

How long does it take to come down the slide?

Use a stopwatch.

Name	Time			
	1st try	2nd try	3rd try	Best try
Tony				
May				
Marlon				

Try sliding down on a plastic shopping carrier canvas shopping bag rubber mat (from the car) straw table mat carpet tile.

Which is fastest/slowest?

Sliding a wooden block

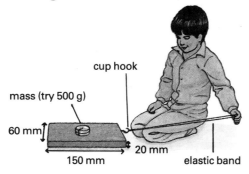

cup hook

mass (try 500 g)

60 mm

150 mm

20 mm

elastic band

Measure the stretch in the elastic band to give the pull needed to start the block moving on different surfaces.

Surface	Pull
Wooden desk top	
Plastic desk top	
Carpet	
Vinyl floor tile	
Playground	

Surface	State	Pull
Spare carpet tile	dry	
	wet	
	oily	
Vinyl floor tile	dry	
	wet	
	oily	

Play shove-halfpenny

Make a board.

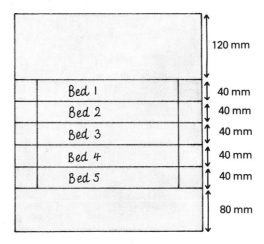

120 mm

Bed 1 — 40 mm
Bed 2 — 40 mm
Bed 3 — 40 mm
Bed 4 — 40 mm
Bed 5 — 40 mm

80 mm

Use 3 × 1p coins
or   3 × 2p coins
or   3 × 5p coins.

First to three coins in each bed wins.

If you play the game with each set of coins in turn, children will get a feel for the different force needed to place each weight of coin accurately.

Try a hovercraft

Make a hovercraft in the classroom. All you need is a lightweight tile and a hairdryer.

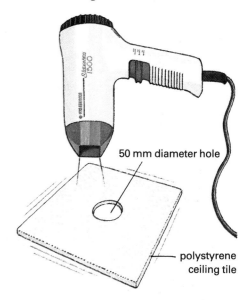

50 mm diameter hole

polystyrene ceiling tile

As air is blown through the hole, it becomes easier to push the tile across a table top. The air forced under the tile reduces the friction between the tile and the table.

Build a light superstructure on your tile so that it looks like a hovercraft.

Principles that children should discover

Friction is the resistance which must be overcome for one surface to slide over another. The smoother the surfaces, the lower the friction.

When one object slides over another, the smoother their surfaces the less friction there is between them.

Moving with sledges

A sledge has runners. Investigate how they work.

How much stretch do you get in the elastic band when you start the sledge moving?

How much stretch do you get in the elastic band when you start the sledge moving on runners?

Improve your technology!

Have a competition to find who can invent the smoothest running sledge. Make sure you give each child the same size body for the sledge in order to keep the test fair.

Name	Types of runner
Emma	Squeezy bottles
Dalvinder	Wooden rulers

Moving with rollers

The Ancient Egyptians probably used rollers to move the large blocks of stone used in building the pyramids.

How much pull is needed to move a brick across the floor?

Now try it using rollers.

Do they help?

Make a self-propelling box using a balloon and plastic drinking straws.

The neck of the balloon should be taken through a small hole in the side of the box.

Try the box and the balloon on the floor without the plastic milk straw rollers.

Moving up ramps

Pull someone along on the flat.

Now pull them up a slope.

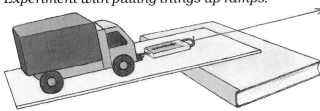

Discuss the nature of the forces needed.

Use a newton meter to measure the force needed to lift a toy lorry vertically.

Experiment with pulling things up ramps.

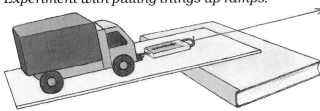

Try varying the height of the ramp.

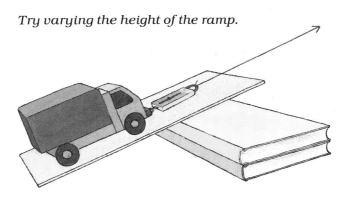

Try varying the length of the ramp, but keep the height constant.

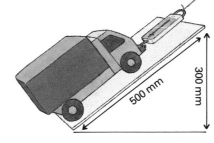

300 mm
500 mm

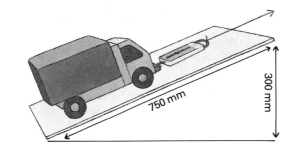

300 mm
750 mm

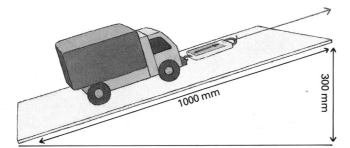

300 mm
1000 mm

Turning a ramp into a screw

Draw a ramp on a piece of paper.

Cut it out and colour along the long edge.

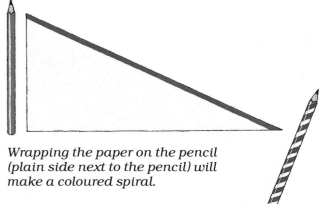

Wrapping the paper on the pencil (plain side next to the pencil) will make a coloured spiral.

A model 'screw'!

Demonstrate to children how a screw can be used to move things.

Make an adjustable seat.

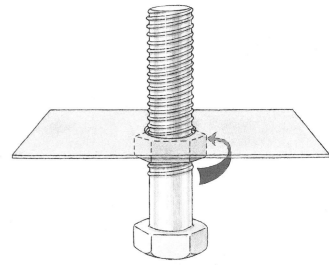

Turning the nut will make the card seat rise.

Collecting, making and improvising 'wheels'

From tin lids.

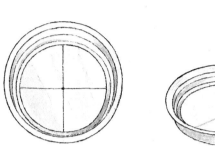

Find the centre point by drawing two diameters.

Punch a hole with a hammer and nail.

nail

plastic bead

From perforated plastic (Gamster) PE balls.

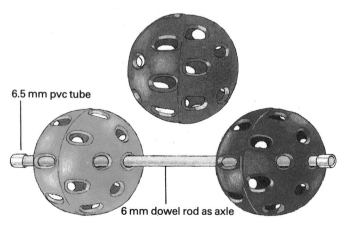

6.5 mm pvc tube

6 mm dowel rod as axle

From squeezy bottles.

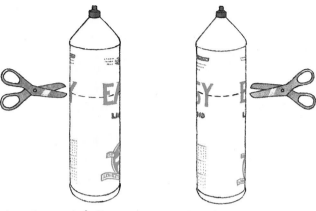

Cut the ends off two squeezy bottles.

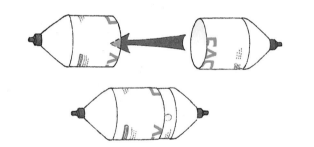

Insert one into the other.

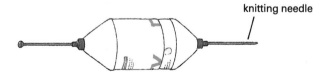

knitting needle

Put a knitting needle through the two ends to act as an axle.

From hobby shops and construction sets: wooden wheels.

From LEGO

Duplo wheels

From Meccano.

From hobby shops: plastic aircraft wheels.

From cotton reels.

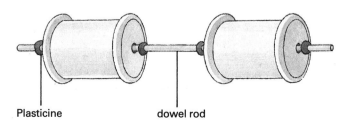

Plasticine

dowel rod

Using the wheels

A trolley.

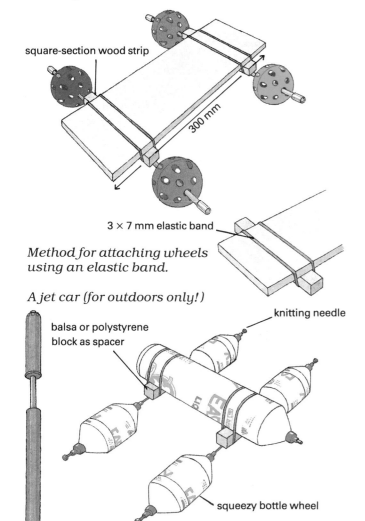

square-section wood strip

300 mm

3 × 7 mm elastic band

Method for attaching wheels using an elastic band.

A jet car (for outdoors only!)

knitting needle

balsa or polystyrene block as spacer

squeezy bottle wheel

Put a mugful of water into the squeezy bottle. Pump in air through the valve. Unscrew the pump and stand clear. Very messy, but great fun! (See also page 15.)

A land yacht.

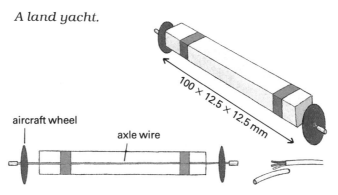

100 × 12.5 × 12.5 mm

aircraft wheel

axle wire

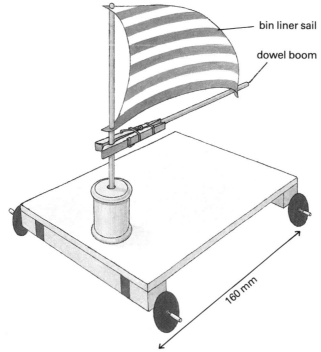

bin liner sail

dowel boom

160 mm

Have a competition. Who can make the land yacht travel furthest? Try different shaped sails on the same base.

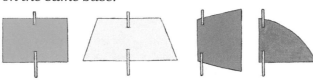

Electric motor power.

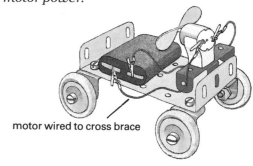

motor wired to cross brace

Elastic band power.

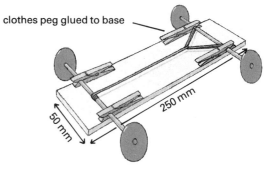

clothes peg glued to base

250 mm

50 mm

Wind the elastic band round the front edge. Then release.

More elastic band power.

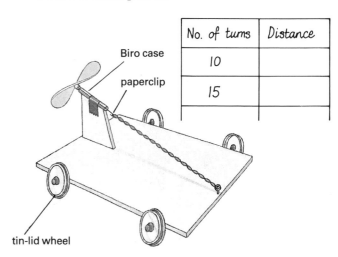

Biro case

paperclip

tin-lid wheel

No. of turns	Distance
10	
15	

Make a large collection of toy cars

Take two cars at a time. Hold them in check with the ruler (starting gate). Let them go.

Which one reaches the floor first?

Which one goes furthest?

Order the collection on distance travelled.

Examine the cars, especially the wheel bearings, and discuss the reasons why children think some perform better than others.

Handicapped cars

From knowledge of their cars' performance ask the children to handicap their cars, i.e. start them from different points to make them all stop at a common finishing line.

Different surfaces

Run the fastest vehicle on to:

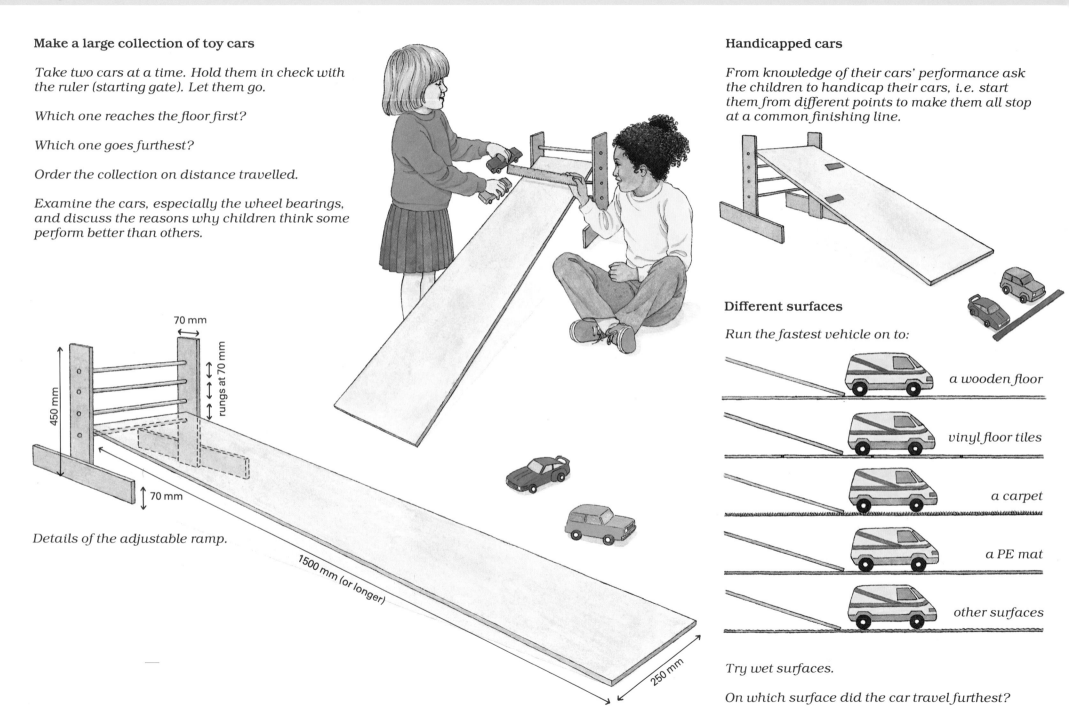

70 mm

rungs at 70 mm

450 mm

70 mm

Details of the adjustable ramp.

1500 mm (or longer)

250 mm

a wooden floor

vinyl floor tiles

a carpet

a PE mat

other surfaces

Try wet surfaces.

On which surface did the car travel furthest?

Varying the incline

Let a lorry run down the slope. Measure how far it travels.

Vary the steepness of the slope.

Slope height	Distance travelled
1	
2	
3	

Varying the load

Keep the height of the slope constant. Vary the load on the lorry (use weights or marbles).

Now vary the height of the slope as well.

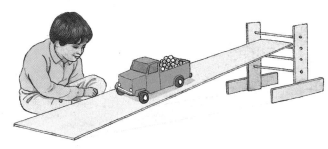

Slope height	Load	Distance travelled
1		
2		

Stability

Badly loaded lorries risk toppling over. Investigate this.

Raise the board.

When does the lorry topple over?

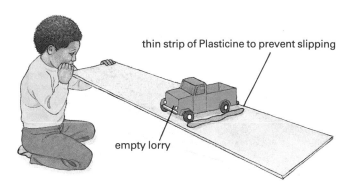

thin strip of Plasticine to prevent slipping

empty lorry

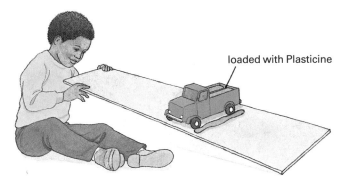

loaded with Plasticine

Experiment with heavier and heavier loads.

Experiment with higher and higher loads.

Questions for discussion

Which surfaces are best for free movement?
Which surfaces would help slow down a vehicle?
How does a slope affect movement?
How does a load affect movement?

Drink cans, squeezy bottles and cotton reels make good rollers.

Drink can dragster

You will need

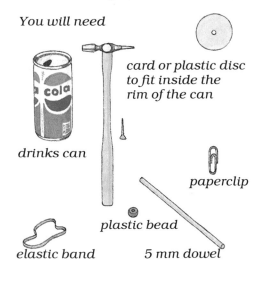

drinks can

card or plastic disc to fit inside the rim of the can

paperclip

plastic bead

elastic band   5 mm dowel

long wire hook
(use a wire coat hanger)

Assemble as follows.

Make a hole through the bottom of the can.

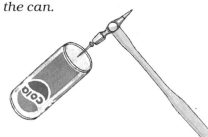

Straighten out the paperclip to make a hook.

Put the hook through the disc and plastic bead.

bead

Bend the straight end of the wire on the dowel rod and twist it around.

Use the long wire hook to pull the elastic band through the can. Secure the other end of the elastic band with a matchstick.

Complete the assembly.

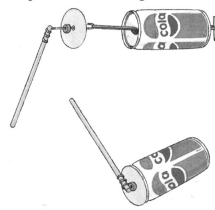

Squeezy bottle roller

You will need

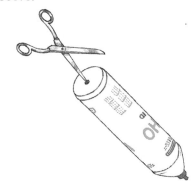

wire coat hanger

squeezy bottle   nail

long elastic band (almost as long as a squeezy bottle)

or

short elastic bands joined together

long wire hook

Cut and bend the coat hanger to make a 'tail'.

Make a hole in the bottom of the squeezy bottle with a pair of scissors.

Pull the elastic band through the squeezy bottle with the wire hook. Secure with a nail.

Replace the hook with the bent coat hanger tail.

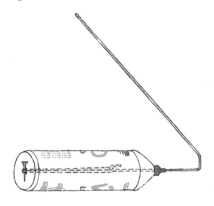

It is essential that this bend is clear of the cap when the band is slack.

Cotton reel tank

You will need

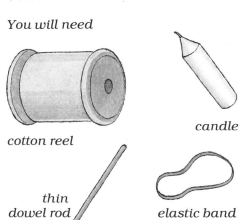

cotton reel

candle

thin dowel rod

elastic band

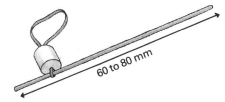

nail or drawing pin

fine wire hook (paperclip)

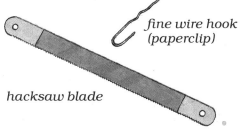

hacksaw blade

Cut a 10 mm wide section from the candle. Use the hacksaw blade with a gentle sawing action.

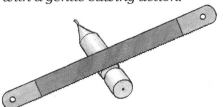

Withdraw the wick to leave a hole.

Thread the elastic band through the candle and secure with the dowel.

 60 to 80 mm

Pull the elastic band through the cotton reel with a hook made from a paperclip.

Secure the free end of the elastic band with a drawing pin. (Thread a nail through if you use a plastic cotton reel.)

Try the rollers

Some testing

Try to improve your models.

What happens when the length of the trailing rod is varied?

What happens when the thickness of the trailing rod is varied?

What happens when the position of the trailing rod is changed?

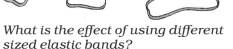

What is the effect of using different sized elastic bands?

What is the effect of varying the number of elastic bands?

What is the effect of varying the size of the roller?

large electrical wire reel

Try a three litre lemonade bottle instead of a squeezy bottle.

Come-back roller

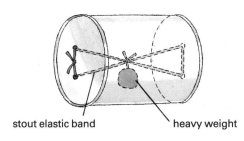

stout elastic band   heavy weight

Roll it away – watch it come back!

Whose dragster will travel furthest?

Should each have the same number of winds?

How will you make the test fair?

How many revolutions to travel 1 metre?

Can you work it out before you test? Use a tape measure to find the circumference of the tin.

For example, a tin with a circumference of 200 mm would take five revolutions to travel 1000 mm (1 metre):

$$\frac{distance}{circumference} =$$

$$\frac{1000}{200} = 5$$

Check.

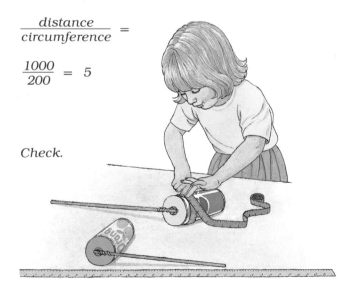

How fast is your dragster?

$$speed = \frac{distance}{time}$$

$$For\ example,\quad speed = \frac{3\ metres}{6\ seconds}$$

$$= 0.5\ metres\ per\ second$$

Does distance covered vary with the number of winds?

Represent your results on a graph.

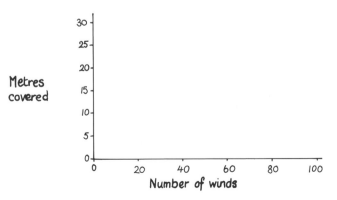

Death-defying leap

Can the dragster leap over the cars?

Try the same tests with the squeezy bottle roller the cotton reel tank.

Make some tracks for your squeezy bottle roller.

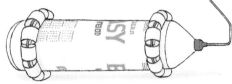

Cut slits into part of the body of a spare squeezy bottle and then turn it inside out.

Make larger wheels using card or polystyrene discs.

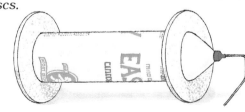

What difference do these modifications make on the results of the tests?

Variations with the squeezy bottle

Pulling a load.

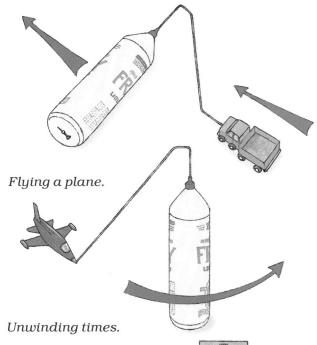

Flying a plane.

Unwinding times.

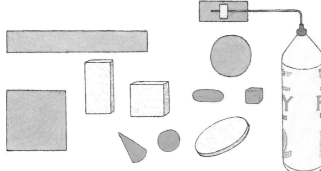

Use shapes from card or polystyrene. Try these flat and edge on.

Use solid shapes made from Plasticine.

Time the different shapes unwinding, each for a fixed number of turns.

Variations with the cotton reel tank

A lifting platform.

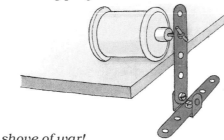

A travelling platform.

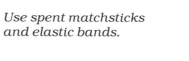

A shove of war!
Who has the most powerful tank?

Good climbers.

Whose tank can climb best? Invent some good climbers.

Wind round with rubber bands.

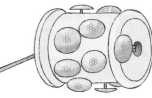

Cover with drawing pins.

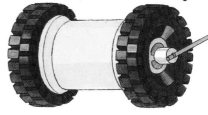

Use spent matchsticks and elastic bands.

Notched wheels.

Fit with toy car tyres.

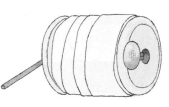

Wind with sticky tape.

The idea of one wheel turning another, either by being just in contact or through some sort of belt drive, is easily demonstrated. Both methods depend on friction.

Wheels in contact

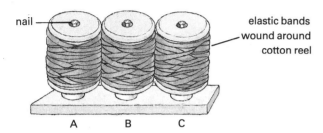

nail —

elastic bands wound around cotton reel

A   B   C

What happens if you turn spool A
   clockwise?
   anti-clockwise?

What happens if you move the other spools?

Try making cogs.

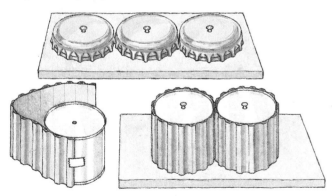

Use pastry cutters to make potato 'gears'.

  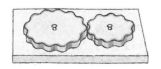

Cut dough gears and bake. Cut clay gears and fire.

Belt drives

Use cotton reels only, first.

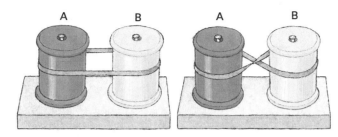

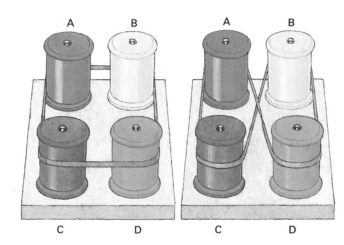

Make sure the reels are free to turn and that the elastic bands are not too tight.

Then use cans as well.

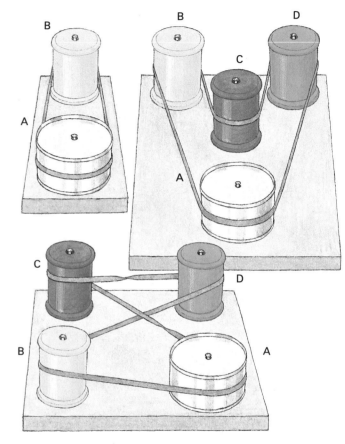

Direction drive wheel A is turned	Direction follower wheels turn	Do the followers go faster or slower?	For one turn of the drive wheel – number of turns or part turns of the followers
	B		
	C		
	D		

Examine a bicycle

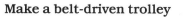

Lift the back wheel. Push on the pedals to make one complete turn of the chain wheel. Count the number of turns made by the back wheel.

Keep a record of your findings.

Number of turns of chain wheel	Number of turns of rear wheel	Number of teeth in chain wheel	Number of teeth in sprocket wheel
1			
2			
3			

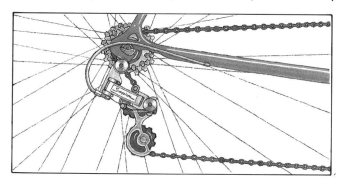

Look at a bicycle with a number of gears. Count the teeth on the chain wheel. Count the teeth on each gear wheel (sprocket).

Find out how many turns each sprocket makes for one turn of the chain wheel.

Make a belt-driven trolley

You will need

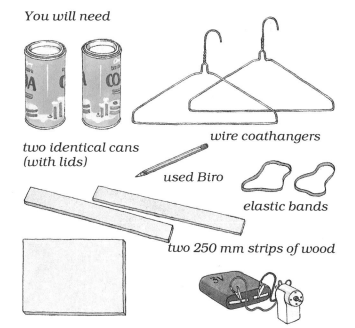

two identical cans (with lids)

used Biro

wire coathangers

elastic bands

two 250 mm strips of wood

piece of wood 40 mm longer than the length of your cans

small battery-driven motor

Make a hole in the centre of each end of each can.

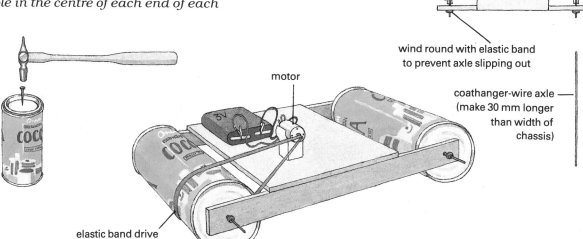

motor

elastic band drive

Drill a hole at each end of your wood strips.

Glue and pin the strips to the main piece of wood.

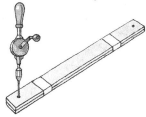

Assemble.

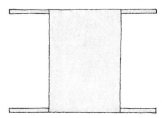

pieces of Biro as spacers

wind round with elastic band to prevent axle slipping out

coathanger-wire axle (make 30 mm longer than width of chassis)

What is the steepest slope the trolley will climb?

A lever is a bar turning about a point.

First-class lever

The turning effect of the *effort* on one side equals the turning effect of the *load* on the other.

load ⟶ □   *effort* ↓

fulcrum ▲

Pushing down on one side makes the other go up.

Second-class lever

The effort and the load are on the same side of the fulcrum, with the load between the effort and the fulcrum.

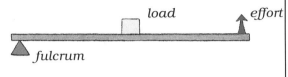

load   *effort* ↑

▲ *fulcrum*

The effort is in the same direction as that in which the load moves.

Third-class lever

The effort and the load are on the same side of the fulcrum but the effort is between the load and the fulcrum. Again the effort is in the same direction as that in which the load moves.

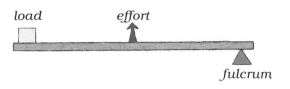

load   *effort* ↑

fulcrum ▲

With this lever a child can move the teacher.

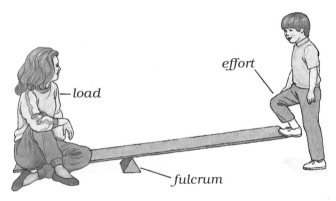

effort

—load

fulcrum

Make a mobile to illustrate this kind of lever.

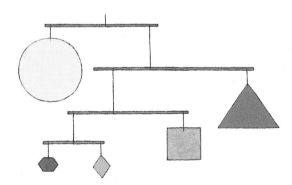

Always begin at the bottom and work upwards.

load   —effort

fulcrum

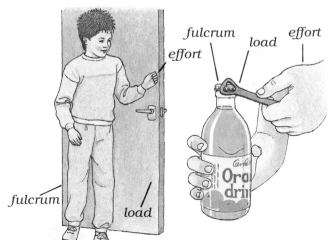

effort   *fulcrum*   load   *effort*

fulcrum   load

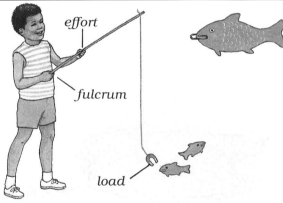

effort   *fulcrum*

load

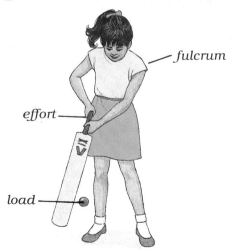

fulcrum

effort—

load—

Make some moving models using levers.

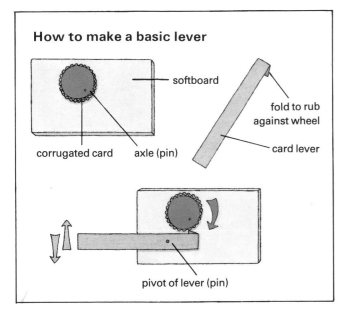

How to make a basic lever

softboard

fold to rub
against wheel

card lever

corrugated card   axle (pin)

pivot of lever (pin)

Horned hat

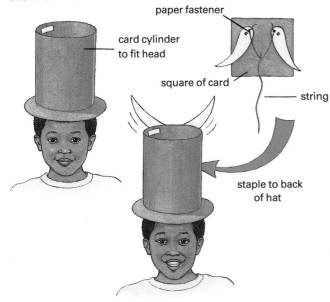

paper fastener

card cylinder
to fit head

square of card

string

staple to back
of hat

Pull string to operate.

Spring-board diver

card fold to rub on wheel

Ocean-going ship

Running mouse

stick

glue

cut legs out
from card

paper fastener

Run the mouse along a table top.

Jumping Jacks

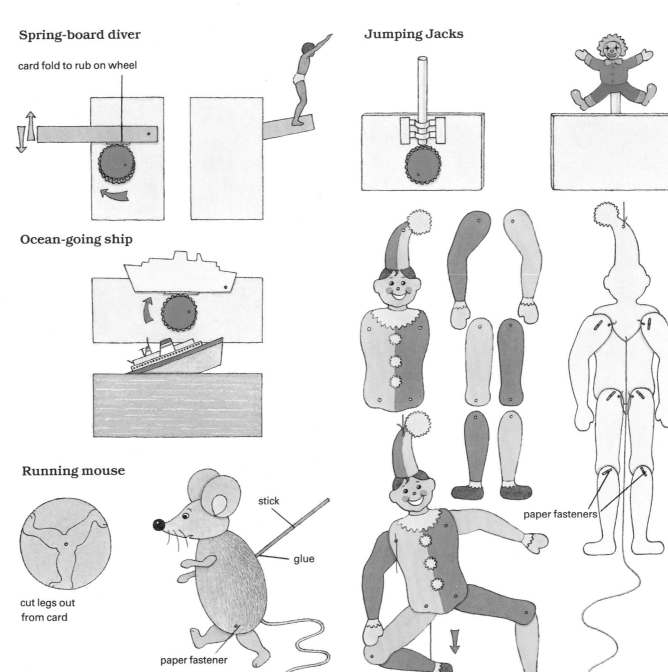

paper fasteners

It is fun to improvise and use home-made pulleys, but these never prove to be very efficient. When you are attempting any work requiring measurement, it will be necessary to use pulleys bought from any of the usual scientific suppliers (see page 94).

However, here are some ideas where improvised pulleys work reasonably well.

Making a basic pulley

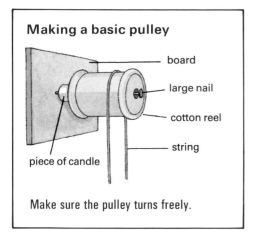

- board
- large nail
- cotton reel
- string
- piece of candle

Make sure the pulley turns freely.

Ski lift

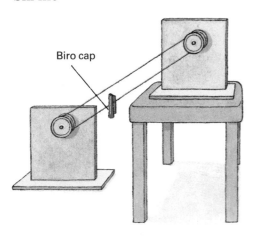

Biro cap

Another improvisation

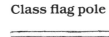

cut and bend

Human pulley

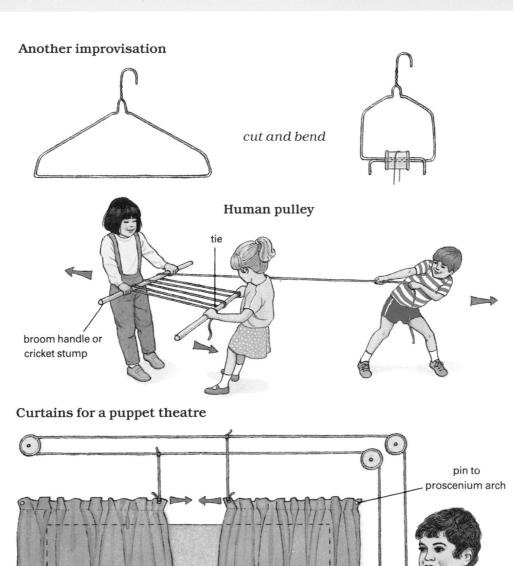

tie

broom handle or cricket stump

Curtains for a puppet theatre

pin to proscenium arch

This is the view behind the stage.

Class flag pole

Using pulleys

Distance brick raised from floor	Distance string pulled

Try different arrangements.

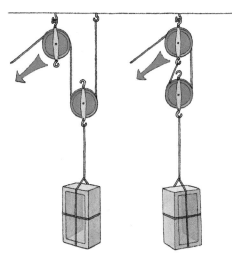

Use a newton meter to measure the force needed to raise the brick.

Record the results.

Distance brick moved	Distance string pulled	Force to lift load

Try arrangements with more pulleys.

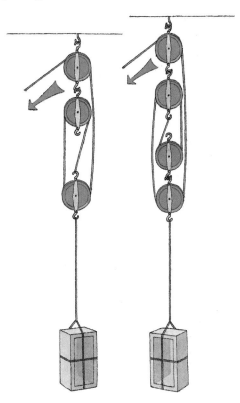

Cranes and pulleys

Make some cranes. Here are two examples.

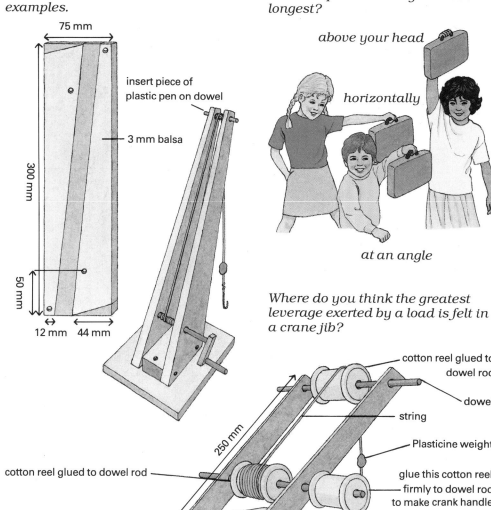

75 mm

insert piece of plastic pen on dowel

3 mm balsa

300 mm

50 mm

12 mm 44 mm

cotton reel glued to dowel rod

dowel

string

Plasticine weight

glue this cotton reel firmly to dowel rod to make crank handle

paperclip hook

250 mm

cotton reel glued to dowel rod

60 mm

200 mm

Hold a school bag in different ways

In which position can you hold it longest?

above your head

horizontally

at an angle

Where do you think the greatest leverage exerted by a load is felt in a crane jib?

Some hull shapes to try.

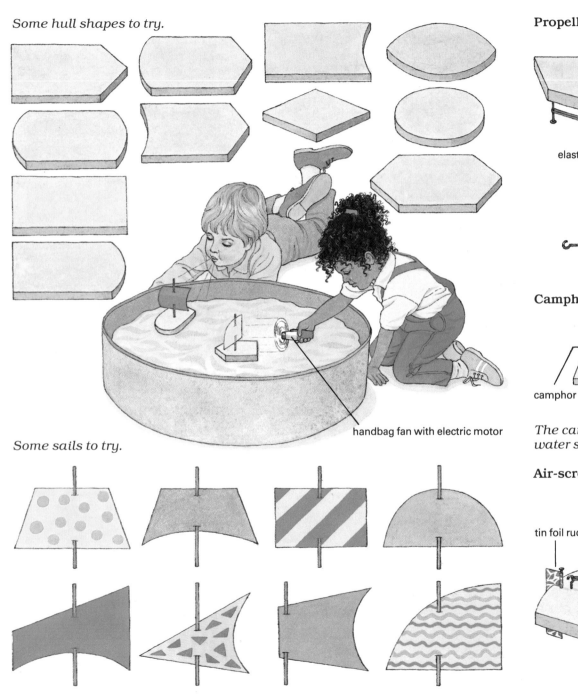

Some sails to try.

handbag fan with electric motor

Propeller-driven boat

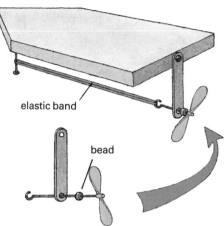

elastic band

bead

Camphor boat

camphor

The camphor must rest on the water surface.

Air-screw boat

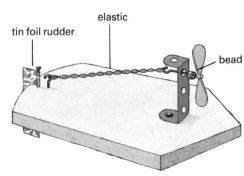

tin foil rudder

elastic

bead

Jet-propelled boat

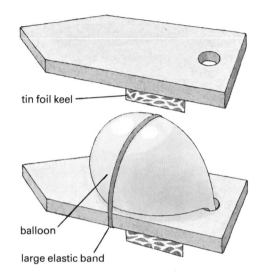

tin foil keel

balloon

large elastic band

Paddle boat

Join pieces together to make paddle.

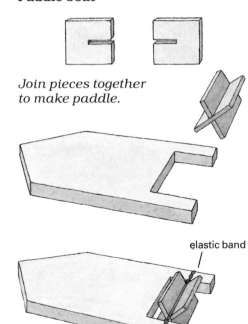

elastic band

Test tank

drop needed so that boat can travel length of tank

at least 2000 mm

150 mm

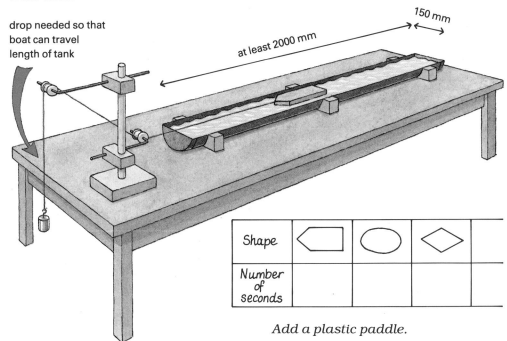

Catamarans

Cut a water softener bottle down the centre.

Connect the two halves by pieces of wood as long as the bottle.

elastic band

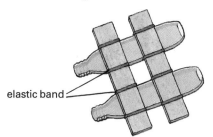

Add a plastic paddle.

Shape			
Number of seconds			

To fix a mast instead, see page 15.

Trimaran

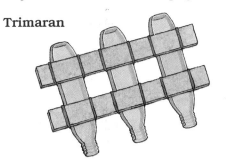

Cargoes

Make some Plasticine boats.

Cut some <u>equal sized</u> pieces of Plasticine.

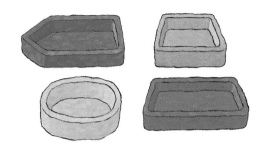

Mould them into boats.

Load each boat with marbles.

Which holds the most?

		Shapes		
○	▭	☐	◁	

Make some aluminium boats.

Make as many different shaped boats as you can using the same amount of aluminium foil.

200 mm

50 mm

Here is one way

Pinch the corners together tightly.

Bend corner round.

Shape the foil.

Make other boats using

100 mm

100 mm

40 mm

250 mm

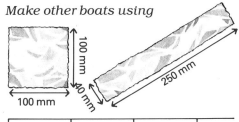

Boat size			
Cargo			

Which carries the most cargo?

Choose a windy day.

Let children run with the wind, and against the wind.

Let them do this again, each child holding a large piece of cardboard (side of a supermarket carton) in front of her/himself.

Experiment with moving at different angles to the wind.

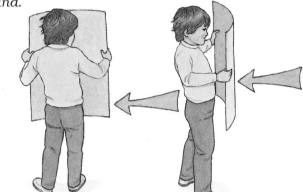

Run 25 m without the cardboard.
Then run 25 m with it.

Name	Time into the wind	
	without card	with card
Julianna		
Billy		

Invent ways of judging wind direction

Drop a handful of grass.

Hold up a wet finger. The wind dries the side of the finger facing it. The finger feels colder on this side.

Watch the direction in which the clouds move in the mirror.

Make a windvane.

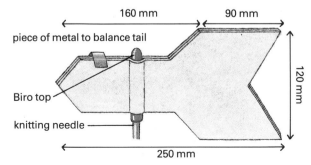

160 mm · 90 mm

piece of metal to balance tail

Biro top

knitting needle

120 mm

250 mm

Measure the strength of the wind

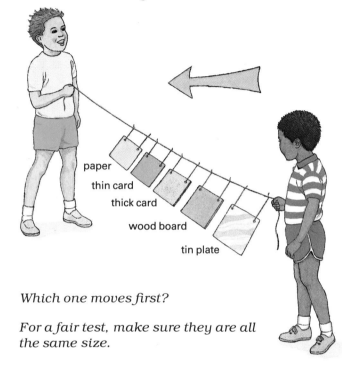

paper
thin card
thick card
wood board
tin plate

Which one moves first?

For a fair test, make sure they are all the same size.

Make a wind-strength meter.

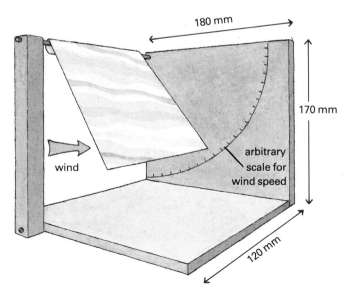

180 mm

170 mm

wind

arbitrary scale for wind speed

120 mm

Kites

One of the simplest, and most effective, kites that children can make is constructed from a dustbin liner and two dowel rods.

Make the kite to any size, but keep to the proportions shown.

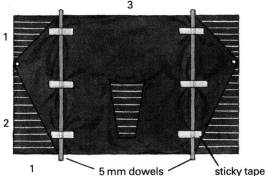

5 mm dowels   sticky tape

Cut away the white-lined portions. Don't forget to cut out the central portion.

Have competitions.

What is the smallest kite you can make that will fly effectively?

Who can make a kite to fly the highest?

Who can lift the heaviest load with a kite?

Wind on wings

Take a sheet of A4 paper and fold in half.

Curve and glue the top half to the bottom half.

Make a hole through both sheets. Insert a length of drinking straw, positioning it between the two holes.

cross-section

Thread a piece of cotton through the straw.

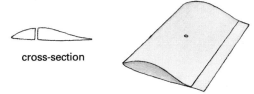

Blow against the front, and then against the rear, of the wing. What happens?

Play 'Flip the kipper'

paper fish

Flying bears

Design a kite to hold a teddy bear in the air for two minutes.

Can you invent a release mechanism to let the bear descend by parachute?

The forces acting on an aircraft are drag, thrust, lift and gravity.

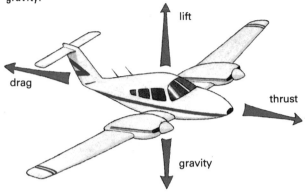

Air is a fluid like all liquids and gases.

Drag

This is the resistance of the air to flight. It acts in the opposite direction to the direction of motion of the aircraft.

Lift

This force acts more or less perpendicularly to the direction of motion. The curved shape of the wing section develops lift.

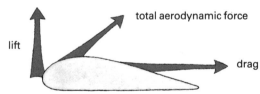

The upper surface of the wing, from its leading edge to its trailing edge, is longer than the lower surface. Air therefore has to travel faster over the upper surface than over the lower surface. This creates a suction above the wing and a pressure below. Together these cause lift.

Thrust

This is provided by the engine.

Make a paper dart

Drag, thrust, lift and gravity can be experienced by children when they fly this paper dart.

1

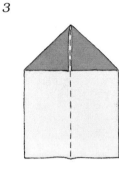

Fold a piece of A4 cartridge paper.

2

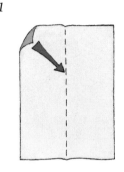

Open out and fold the corner over.

3

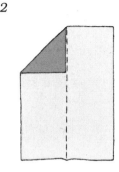

Fold the other corner over.

4

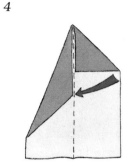

Fold again, like this.

5

Turn the other corner down.

6

Turn over.

7

Fold sides to centre.

8

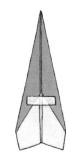

Fold again.

9

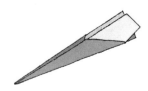

Fold in half.

10

Hold the centre fold and open out.

11

Try your dart out.

Improving the design

Get a good balance.

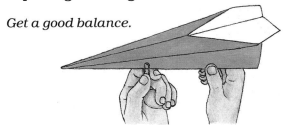

Use a paperclip to ensure that the dart balances <u>*about its centre.*</u>

Does this improve flight?

Add tail flaps.

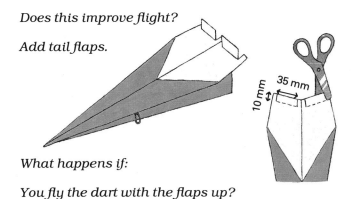

35 mm
10 mm

What happens if:

You fly the dart with the flaps up?
You fly the dart with the flaps down?

Add side stabilizers.

70 mm

What effect do the stabilizers give?

Up?

Down?

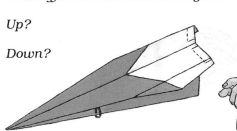

Experiment with different sized darts.

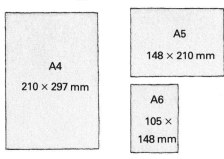

A4
210 × 297 mm

A5
148 × 210 mm

A6
105 × 148 mm

Size	Distance			
	Try 1	Try 2	Try 3	Best
A4				
A5				
A6				

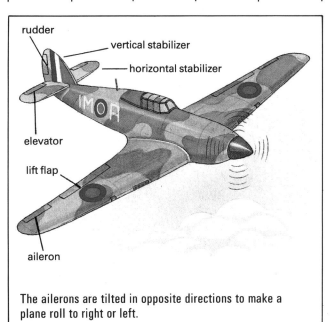

rudder

vertical stabilizer

horizontal stabilizer

elevator

lift flap

aileron

The ailerons are tilted in opposite directions to make a plane roll to right or left.

Turn the dart into a delta wing craft

Add an extra wing to the dart like this.

Fold a sheet of A4 cartridge paper in half.

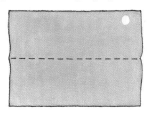

Open out and fold in the corners.

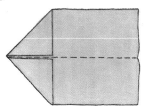

Fold again.

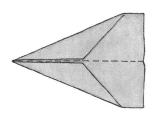

Fit the dart into this extra wing. Stick with a touch of glue.

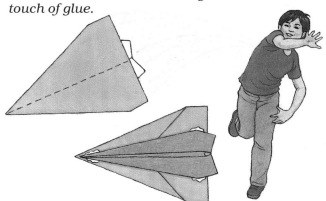

Acrobatic plane

Fold a sheet of A4 paper crosswise.

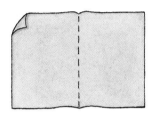

Open up and fold a crease 13 mm from the edge of the long side.

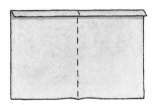

Fold and fold again several times.

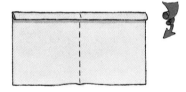

Refold to centre line. Cut out a notch in the paper as below.

Open out.

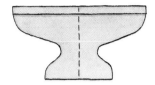

Fold the wing tips <u>up</u>.

Fold <u>down</u> the outer edges of the tail.

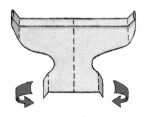

Check the plane for symmetry. It is essential that one half is a mirror image of the other. If you have any imbalance take a fresh piece of paper and begin again.

Try the plane.

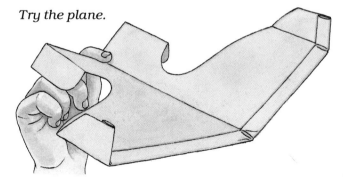

Launch it gently away from you with a slight downward motion.

It should fly like this.

Tips

If the plane glides all right but moves from side to side, check the symmetry again.

If the plane dives, turn the trailing edge of the tail up a little.

If the plane undulates, turn the trailing edges of the tail down. If this is not successful, the plane may be too heavy in the tail. Try a paperclip or tiny piece of Plasticine on the nose.

These variables will affect performance

The speed of launching.

The amount of up-elevation provided by the tail.

The angle at which the plane is launched.

Experiment with these to try for some loops. Don't be afraid to try plenty of up-elevation.

Balsa wood glider

Cut out the wing. Score and crack at the centre.

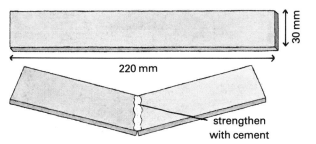

220 mm

30 mm

strengthen with cement

Make the tail and tail fin. Then glue the pieces together.

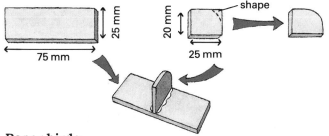

25 mm

75 mm

20 mm

25 mm

shape

Cut out the fuselage.

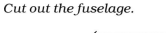

160 mm

20 mm

3 mm from top

25 mm

Glue wings into the fuselage.
Glue tail and tail fin to rear tip of the fuselage.

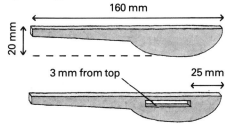

Jet rocket

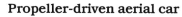

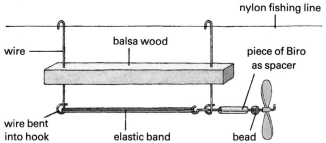

nylon fishing line

two short pieces of plastic drinking straw

balloon

sticky tape

bulldog clip to hold air until ready to launch

Propeller-driven aerial car

nylon fishing line

wire

balsa wood

piece of Biro as spacer

wire bent into hook

elastic band

bead

How far will each vehicle travel?
What happens if you vary the amount of air, or the number of twists of the elastic band?

Paper birds

Use this template to make a paper bird.

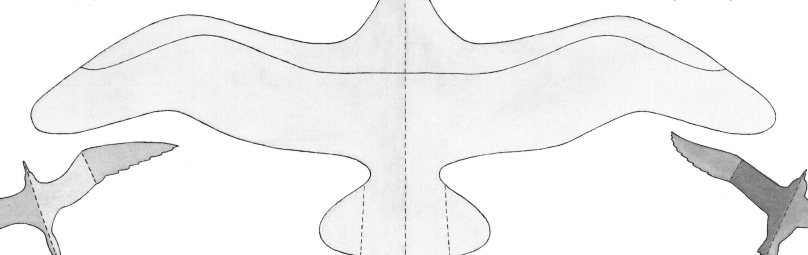

Parachutes

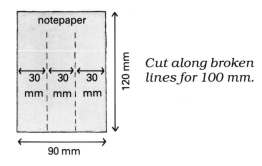

Number of washers	Time to fall
1	
2	
3	

Make a paperclip hook.

cut

Assemble the parachute.

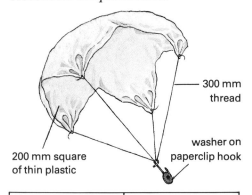

300 mm thread

200 mm square of thin plastic

washer on paperclip hook

Parachute size	Time to fall
200 mm square	
300 mm square	

Try parachutes of different materials: plastic, cotton, nylon, paper.

Try parachutes with a hole at the centre of the canopy.

Do different sized holes make a difference?

Have a competition to find who can land a toy vehicle most gently.

Spinners

notepaper

30 mm | 30 mm | 30 mm

90 mm

120 mm

Cut along broken lines for 100 mm.

Fold to make the spinner.

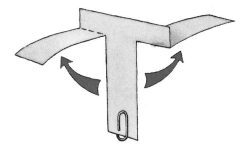

Investigate the variables.

Change the size of the flaps.

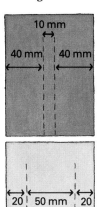

10 mm

40 mm | 40 mm

20 mm | 50 mm | 20 mm

Make different sized spinners.

Try halving the size. Then try doubling, quadrupling . . . it.

Try different materials: tissue paper, newspaper, card, sugar paper.

Investigate reversing the flaps.

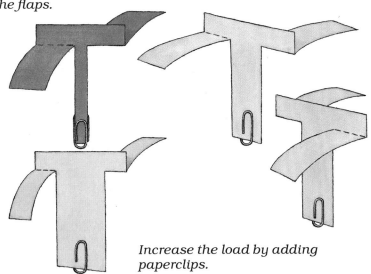

Increase the load by adding paperclips.

Boomerangs

Cut out a boomerang from card.

Use this template.

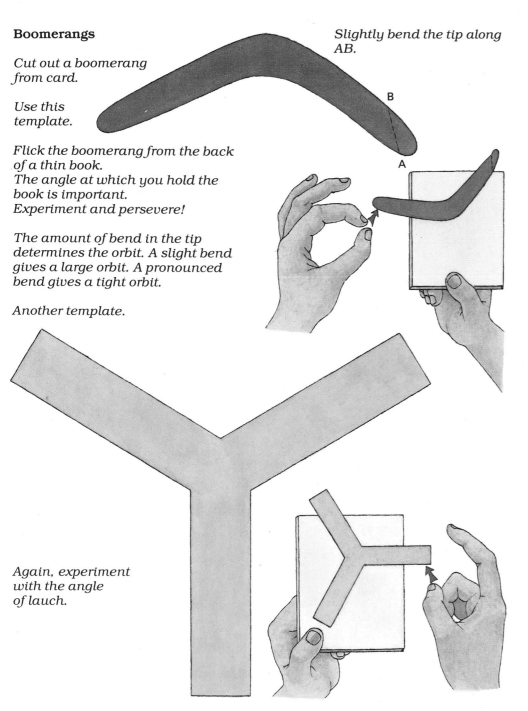

Flick the boomerang from the back of a thin book.
The angle at which you hold the book is important.
Experiment and persevere!

The amount of bend in the tip determines the orbit. A slight bend gives a large orbit. A pronounced bend gives a tight orbit.

Another template.

Again, experiment with the angle of lauch.

Slightly bend the tip along AB.

B

A

Tops

Collect tops and compare them as spinners.

Make some tops from card. Cut out circles and use dowels or pencils as spindles.

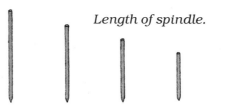

Investigate the variables.

Length of spindle.

Size of card disc.

Thickness of card.

Position of the spindle.

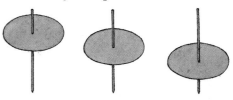

Try different layers.

Investigate other spinners such as a gyroscope.

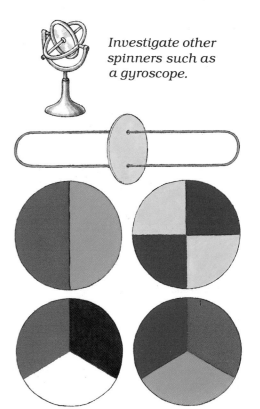

Hot air balloon

1

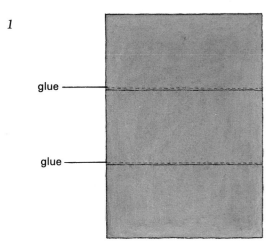

glue

glue

Glue three <u>large</u> sheets of tissue paper together. Do this six times.

2

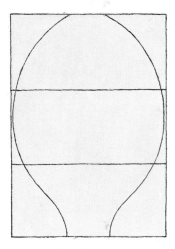

Pile the six sheets on top of each other and draw a balloon panel on the top one. Cut them out.

3

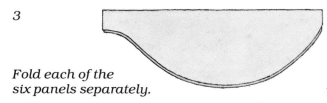

Fold each of the six panels separately.

4

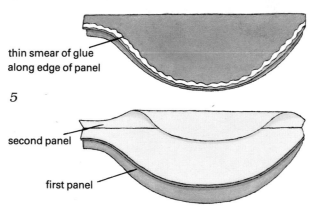

thin smear of glue along edge of panel

5

second panel

first panel

Glue the first panel to the second with a thin smear of PVA glue, about 10 mm from the edge.

6 *Keep this process up until all six sections are joined.*

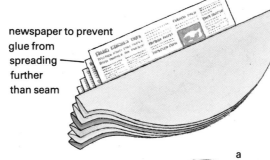

newspaper to prevent glue from spreading further than seam

7

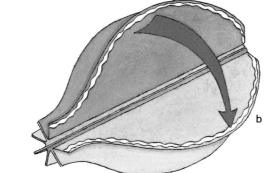

a

b

Glue a to b. Leave to dry.

8 *Shape up the balloon. Fold a sheet of tissue in four to make a collar. Add a strengthening disc of tissue to the top.*

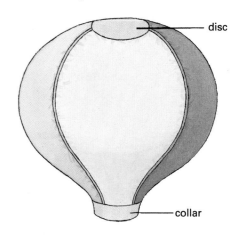

disc

collar

9 *Fly the balloon. Use hot air from a hairdryer or a convector heater to inflate the balloon.*

Experiment with balloons of different sizes.

Measure the maximum height the balloon reaches using a clinometer. (See 'An Early Start to Nature', page 9).

Balloons can also be made of plastic. The lightweight plastic used to cover clothes returned from the dry cleaners is ideal.

Spinning propeller

You will need

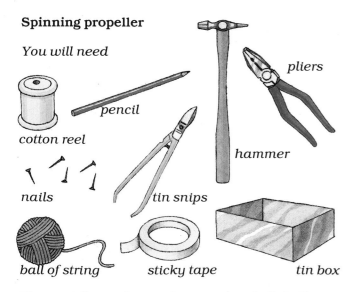

cotton reel

pencil

pliers

hammer

nails

tin snips

ball of string

sticky tape

tin box

Hammer the nails into the cotton reel. Cut off their tops with a pair of pliers.

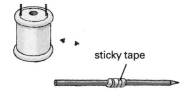

sticky tape

Wind the sticky tape round and round the pencil, enough for the cotton reel to sit on.

Cut a piece of tin to shape. Make two holes in it, the same distance apart as the two nails.

dull edge for safety

twist

Sit the propeller on the nails.

Wind string round the cotton reel. Sit it on the pencil. Revolve quickly to spin the propeller.

Another model

You will need

5 mm dowel rod

flat sided plastic container

Cut out a strip from the container.

100 mm

20 mm

Using a nail, make a hole through its centre. The diameter must be less than the dowel's.

Push the stick through the hole. Twist the strip slightly.

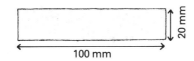

Spin the helicopter in your hand and throw it in the air.

Explore the variables

Vary the length of the blade.

Vary the width of the blade.

Try using two, three, four blades at angles to each other.

Vary the position of the blade along the rod.

Toy windmill

pinhole

100 mm

100 mm

cuts

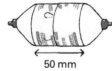

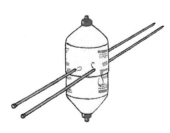

Try with pinholes in these positions too.

Wind vane

You will need

tape

3 long knitting needles

4 squeezy bottles

Cut the tops off three of the bottles, 50 mm below the rim.

Push one top right inside the other to make a hub.

50 mm

Push two knitting needles through the hub an *equal* distance apart.

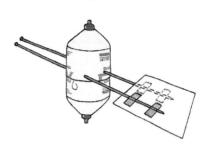

Cut two rectangular blades of plastic 100 × 50 mm from the unused squeezy bottles.

Fit the blades to the knitting needles with tape.

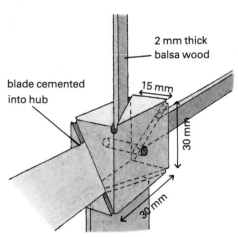

The blade fits <u>over</u> one needle and <u>under</u> the other.

Assemble the vane.

bend

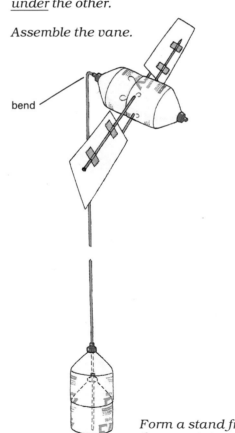

Form a stand from the remaining tops.

Windmill model

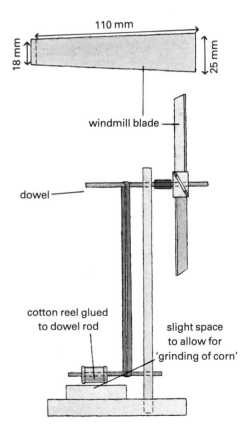

2 mm thick balsa wood

blade cemented into hub

15 mm

30 mm

30 mm

110 mm

18 mm

25 mm

windmill blade

dowel

cotton reel glued to dowel rod

slight space to allow for 'grinding of corn'

Water wheel

You will need

2 litre lemonade bottle

knitting needle

hacksaw blade

straight-sided cork

Cut a large window in the side of the bottle. Make two opposing holes with the knitting needle.

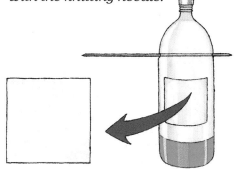

Cut strips from this window for the water wheel. (Use plastic from a second bottle, if necessary.)

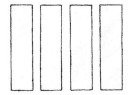

Cut four slits in the cork.

Push the knitting needle through the cork to make a hole.

Insert the blades and pull out the needle.

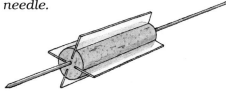

Insert the wheel into the bottle through the window. Rethread the needle.

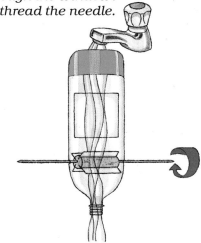

Can you use this wheel to turn other things?

Steam turbine

You will need

large tin can

hacksaw blade

aluminium kitchen foil

straight-sided cork

knitting needle

tape

2 Meccano strips

Cut out a circle of foil (the size will depend on the size of the tin used).

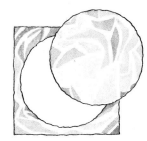

Cut slits in the foil.

Cut two discs of cork.

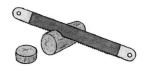

Glue the cork discs each side of the foil. Bend the foil to make blades.

Make a hole in the lid of the tin.

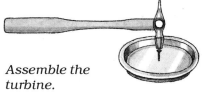

Assemble the turbine.

Take care!

heat source

Children are surrounded by things made of stone, concrete, wood, plastic, metals, artificial and natural fibres and so on. Some are featured in certain songs of childhood.

London Bridge is falling down

1 London Bridge is falling down, etc.

2 How shall we build it up again? etc.

3 Build it up with wood and clay, etc.

4 Wood and clay will wash away, etc.

5 Build it up with iron and steel, etc.

6 Iron and steel will bend and bow, etc.

7 Build it up with gravel and stone, etc.

8 Gravel and stone will fall away, etc.

9 London Bridge is falling down, etc.

The Carpenters Song

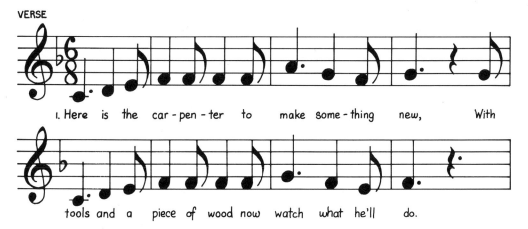

VERSE

1. Here is the car-pen-ter to make some-thing new, With tools and a piece of wood now watch what he'll do.

CHORUS

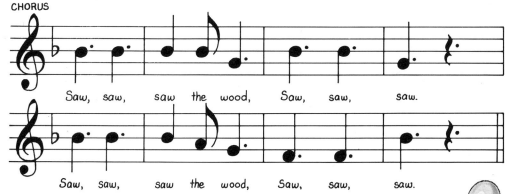

Saw, saw, saw the wood, Saw, saw, saw.

Saw, saw, saw the wood, Saw, saw, saw.

2 Six bits of wood he's sawn, but each one is rough,
He'll use a plane on every piece until it's smooth enough.
Smooth, smooth, smooth the wood,
Smooth, smooth, smooth.

3 Now he's got a compass and a circle he can draw.
And then he'll use a brace and bit, and slowly start to bore.
Bore, bore, bore a hole.
Bore, bore, bore.

4 Through the hole that's in the wood he puts a special blade,
To saw round the circle, so a bigger hole is made.
Saw, saw, saw the wood,
Saw, saw, saw.

Making a stringed instrument

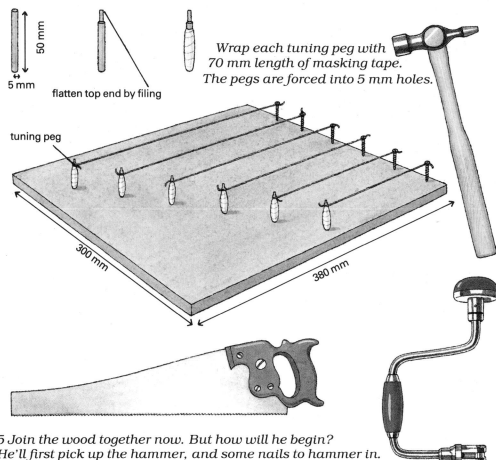

50 mm

5 mm

flatten top end by filing

Wrap each tuning peg with 70 mm length of masking tape. The pegs are forced into 5 mm holes.

tuning peg

300 mm

380 mm

5 Join the wood together now. But how will he begin?
He'll first pick up the hammer, and some nails to hammer in.
Hammer, hammer, hammer the nails,
Hammer, hammer, hammer.

6 Fix to the wooden box some pegs and some screws,
A screwdriver is the tool that he'll have to use.
Turn, turn, turn the screws,
Turn, turn, turn.

7 Strings go from screw to peg, around and around,
Then turn the peg to tune the strings and listen to the sound.

(Zither plays final chorus.)

How can we keep ourselves warm?

Try putting on several layers of clothes.

Wrap a child in a blanket.

Wrap a child in newspaper.

Make a graph of coats worn to school on a winter's day.

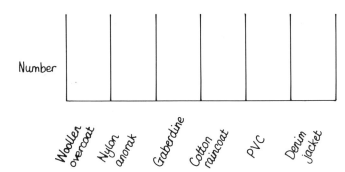

Number

Woollen overcoat — Nylon anorak — Gaberdine — Cotton raincoat — PVC — Denim jacket

How can we keep our feet warm?

Wear Wellington boots. Insulate one foot with various materials, keep the other bare as a control.

Try

sock

newspaper

plastic shopping bag

tea towel

kitchen foil

sugar paper

hay or straw

shredded paper

How can we keep our hands warm?

Try a variety of materials.

Can you knit a hat to keep your head warm?

Which kind of stitch is best for a tight fit?

See page 69 for making a bobble.

Conductors and insulators

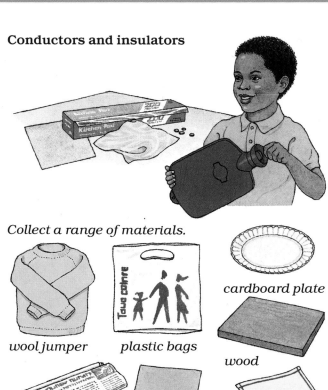

Collect a range of materials.

wool jumper

plastic bags

cardboard plate

wood

newspaper

writing paper

cotton handkerchief

coins

kitchen foil

china plate

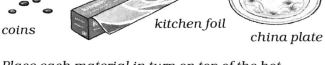

Place each material in turn on top of the hot water bottle. Leave for a few minutes.

Which materials feel warmest?

Which materials feel coldest?

Which would make the best insulators for keeping warm?

Use thermometers to take the temperature in a series of mugs filled with hot water.

Same materials, different shapes

Different materials

enamel   plastic   china   pottery

What observations do children make from feeling the mugs as they cool, and from reading the thermometers?

Plastic cups inside one another

1 cup   2 cups   3 cups   4 cups

Try insulation

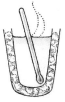

newspaper   cotton wool   kitchen foil

Which teapot keeps the warmest longest?

Try and match them for capacity before you test.

pottery   metal   china

100°c

50°c

Temperature

Time in minutes

Compare a large pot with a small pot

100°c

50°c

Temperature

Time in minutes

Does a tea cosy help? Keep the capacity of the pots the same. Vary the cosies.

fur muff   kitchen foil   tea towel

Which 'bear' keeps the warmest longest?

Fill each bear with hot water.

father bear

Take care!

baby bear

mother bear

What effects do
*   size*
*   shape*
*   volume*
have on cooling?

Who is the coolest?

Do some dressing up – and dressing down!

Which ice cube takes longest to melt?

Make some identically sized ice cubes. Put each on a different coloured card background. Place in the sun.

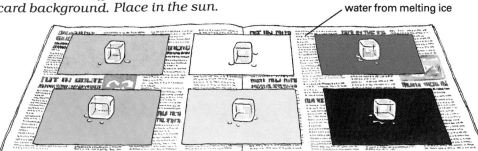

newspaper to absorb water from melting ice

Which colour absorbs heat the quickest?

Which colour would be best to help keep ice cubes?

Ice cube competition

Give an ice cube to each child. Who can keep it longest?

Provide lots of materials that might act as insulators:
*   newspapers*
*   kitchen foil*
*   old hankies*
*   cotton wool*
*   shredded paper*
*   plastic bags.*

Some children may think of putting their ice cubes inside a container insulated inside another container.

Invent devices for keeping bottled milk cool

cut away 2 litre lemonade bottle

water

ice cubes

aluminium foil

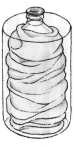

wet cloth

air alone

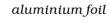

polystyrene chippings

Is it best if the containers have lids?

Whose device is best? How will you make the test fair?

Keeping things dry

Invent ways of waterproofing fabrics.

Test a range of fabrics to find which are waterproof.

Keep fabric taut during the test.

Try rubbing on

candlewax

polish

glue

nail varnish

Test for waterproofing.

Keeping things wet

House plants need to be kept watered when we are on holiday.

Invent devices for doing this.

Make a collection of fabrics.

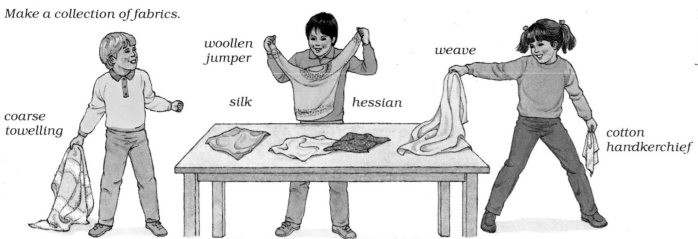

woollen jumper

weave

silk

hessian

coarse towelling

cotton handkerchief

Investigate the structure of different fabrics

Look at each type of fabric under a binocular microscope (or hand lens).

Make drawings to show how each fabric is woven.

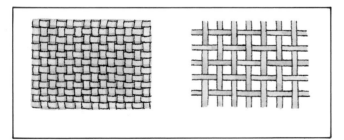

Cut a strip from each piece of fabric.

10 mm

How many threads can you tease out from each one? Look at these different threads.

Are they made up of one fibre, or of several twisted together?

Make drawings of what you can see through a microscope.

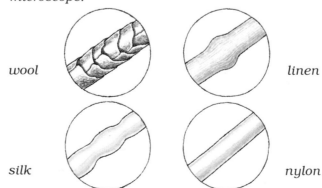

wool

linen

silk

nylon

Do the same with different sewing threads taken from cotton reels.

Investigate the flammability of different fabrics

Burn a short length of thread from each type of fabric.

Take care!

Keep a record of your findings.

Does it burn or melt?
Does it burn slowly, or flare up?
Does it smoke?
Does it smell?

Can the fabrics be made flame resistant? Soak the fabric that you have found to burn easily in various solutions:

salt   borax   alum   borax and alum

Let the pieces dry, then try burning them again.

Salt – sodium chloride
Borax – sodium borate
Alum – potassium aluminium sulphate

Investigate the strength of different fabrics

How hardwearing is each fabric?

Take a range through from a cotton dress, to school uniform, jeans, and overalls.

Test the least worn part of used garments.

Record your results.

Number of rubs	Comment
10	
20	
30	
40	

Which fabrics tear easily?

Use a rod with a nail through its end.

Which threads and fibres stretch most?

Hang some over night.

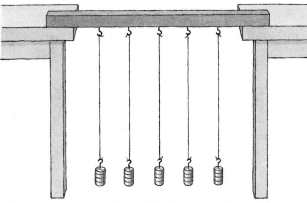

Make sure each thread is the same length to begin with.

Make sure each thread carries the same load.

Invent ways of testing the strength of thread and fibres.

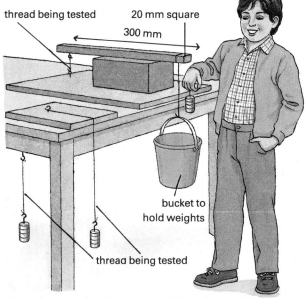

thread being tested   20 mm square

300 mm

bucket to hold weights

thread being tested

Are the results the same if the threads or fibres are wet?

Investigate the stain resistance of different fabrics

Which fabrics protect from dirt and mess? Try dropping on oil, ink, powder paint, grease. Which stain the white paper?

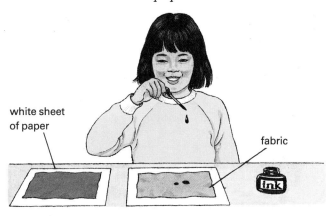

white sheet of paper

fabric

ink

One of the oldest technological activities known to us is that of weaving.

Try some simple weaving.

Card loom

1 Cut notches in a piece of thick card.

2 Wind wool fairly tightly around the card. Hold the lead end with sticky tape.

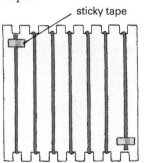

sticky tape

3 With a large darning needle, weave a piece of different coloured wool.

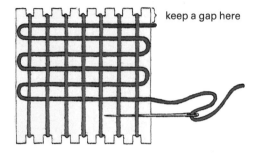

keep a gap here

4 Cut the loops of wool which go through the notches.

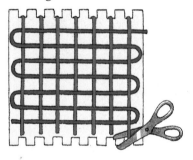

Wooden loom

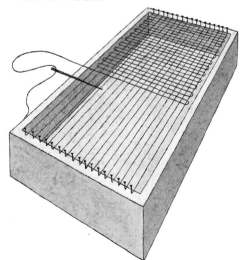

Paper weaving

Fold a sheet of paper in half and make a series of equidistant cuts.

Open it out and weave in strips of paper.

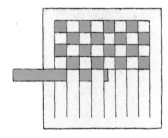

Weave different patterns.

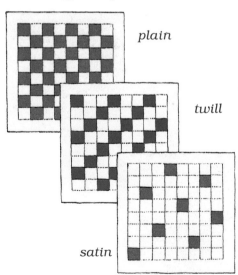

plain

twill

satin

Bridget Cross

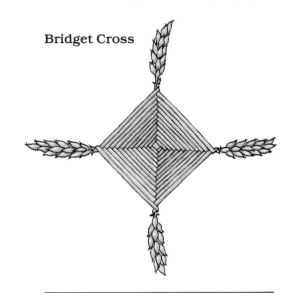

A wheat plant

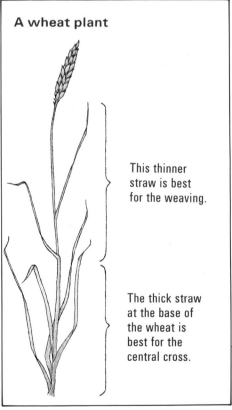

This thinner straw is best for the weaving.

The thick straw at the base of the wheat is best for the central cross.

1 Form a central cross from two thick pieces of straw.

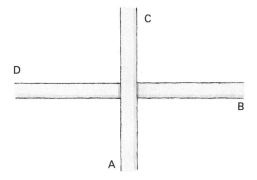

2 Start weaving like this.

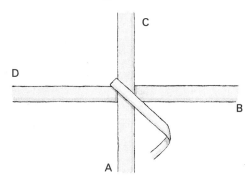

3 Take the weaver straw under A and over the centre to bend down behind B.

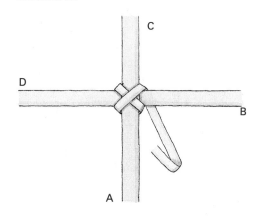

4 Wind the straw back across the centre. Bend it around the back of C, over the front and up behind D.

5 Add new weaver straws as you go along by inserting the thin end of the new straw into the thick end of the old weaver straw.

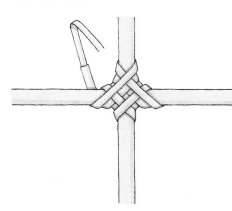

6 Continue until the cross is finished. Tie off the final end with a piece of thread.

Cut four heads of wheat, each with a 100 mm stem.
Insert the stems in the corners of the cross.

Some 3D shapes

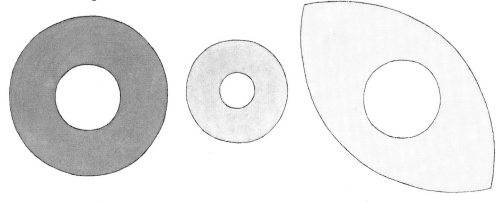

Choose one of these shapes.

Cut two copies of it from card.

Remove the central portion in each case.

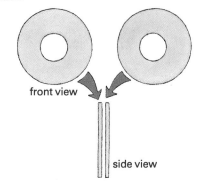

front view

side view

Join the two shapes together by taking a piece of wool through the central hole, then round the two card rings and back through the central hole.

Wind the wool firmly and evenly. Add different colours as you go along.

When fully wound cut along between the two cards and tie a piece of wool tightly around the centre.

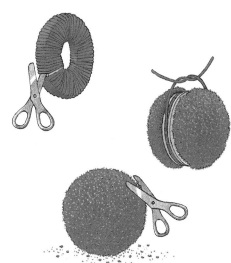

Remove the cards.

A pom-pom will emerge. Trim any loose pieces.

Various parts of plants such as red cabbage, beetroot, peapods, lichens, and so on, can be used to provide dyes. Instant coffee and tea can be used too.

Mordants

Natural dyes need a mordant. This is a chemical which fixes the dye to the fibres. (Latin 'mordere' – 'to bite'.)

Some common mordants are:

alum – potassium aluminium
     sulphate
iron – ferrous sulphate
chrome – potassium dichromate

Mordanting

Mix the mordant with water to make a solution.

Alum

25 g to each 100 g of material.
1 teaspoon of cream of tartar (for extra brightness).
Simmer.

Chrome

10 g to each 450 g of material.
Keep the lid on since chrome is light-sensitive.

Iron

20 g to each
100 g of material.

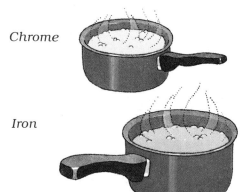

Some natural dyes

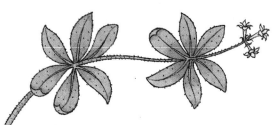

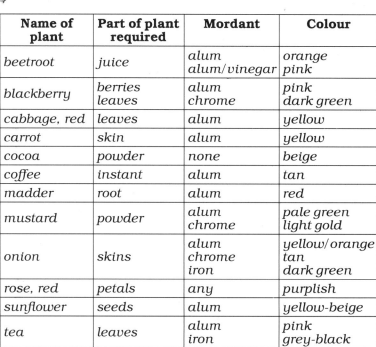

Name of plant	Part of plant required	Mordant	Colour
beetroot	juice	alum alum/vinegar	orange pink
blackberry	berries leaves	alum chrome	pink dark green
cabbage, red	leaves	alum	yellow
carrot	skin	alum	yellow
cocoa	powder	none	beige
coffee	instant	alum	tan
madder	root	alum	red
mustard	powder	alum chrome	pale green light gold
onion	skins	alum chrome iron	yellow/orange tan dark green
rose, red	petals	any	purplish
sunflower	seeds	alum	yellow-beige
tea	leaves	alum iron	pink grey-black

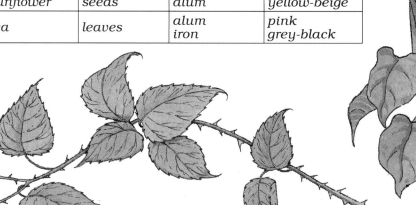

Chop up leaves, roots, skins or whatever. Crush berries.

Soak with water overnight.

Then boil for 20 minutes.

Cool then strain.

Cover material with dye solution. Simmer for 15 minutes. Stir occasionally. <u>Don't</u> use a stained stick from a previous dyeing.

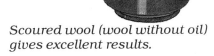

Scoured wool (wool without oil) gives excellent results.

When using a mordant, you need to soak the material (cotton, wool, linen) before adding the dye. See how materials such as nylon and rayon take up dyes.

Comparing dyes

It is said that a good dye must

give a good strong colour
dissolve in water (for dyeing natural materials)
be easily absorbed by fibres
stand up to washing
not fade in the light.

Extract some of the natural dyes and test them out against the above criteria.

Sorting variables

What effect does each of the following have

dyeing directly without a mordant
using a mordant first
varying the dyeing time
varying the temperature
dyeing while simmering
dyeing while cold
dyeing in tap water/dyeing in distilled water?

Testing for fastness

Heat the sample in soapy water for 15 minutes.

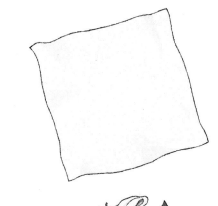

Tie-dyeing

Tie fabrics in various ways and dye.

The dye cannot get at the tied parts and this produces patterns.

Batik (wax resist dyeing)

Heat a wax mixture (30% beeswax, 70% paraffin wax) at about 50°C. Paint this on your design on the fabric. The waxed parts prevent the dye getting at the fabric.

Remove the wax by boiling in water and detergent after dyeing.

Try Dylon dyes for tie-dyeing and batik.

Obtain a set of metal discs from one of the laboratory suppliers.

| copper | lead | steel | aluminium |

| brass | iron | zinc | tin plate |

Carry out tests on each metal

How does it feel?

What colour is it?

Does it shine?

Keep a record of your results.

Is it heavy or light?

Can you put the discs in order of heaviness?

Is it hard?

Drop a small ball bearing down a plastic tube (made from rolled sheet).

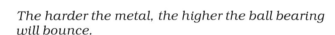

The harder the metal, the higher the ball bearing will bounce.

Will it bend?

Use your fingers.

Will a magnet attract it?

Will electricity pass through it?

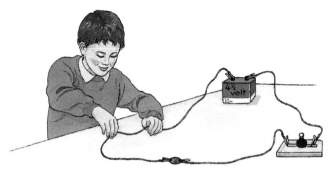

Does it float?

What kind of sound does it make if dropped on a paving stone?

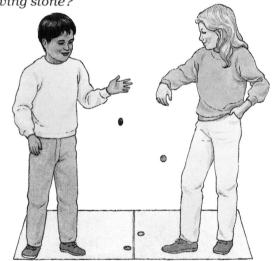

Metal	Feel?	Colour?	Does it shine?	Heavy / Light?	How hard?	Does it bend?	Is it magnetic?	Does it conduct electricity?	Does it float?	What sound does it make?

Tarnishing

Collect examples of different metals.

Clean them with emery paper. Rub well.

Metal	Colour of metal before cleaning	Colour of metal after cleaning

Put them out of doors.

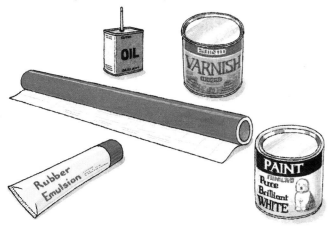

Invent ways of protecting the metals.

Try chemical plating

Put a clean piece of iron in copper sulphate solution.

The jar must be clean.

Try electroplating

Use a nickel sulphate solution with nickel and copper electrodes.

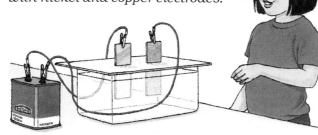

Make a solar cooker

Here the shininess of some metals is put to use.

Cut the front from a strong shoe box. Make two holes in the sides.

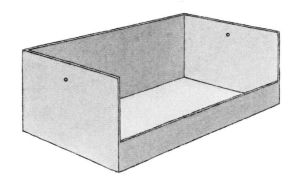

Cut a disc from card.

Its radius needs to be less than the depth of your box.

Cut the disc in half.

Use the two semicircles and another piece of card to make this round structure.

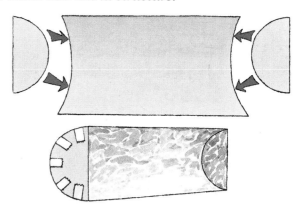

Line the interior with aluminium foil. You now have a mirror to focus the sun's heat.

Fix it to the shoe box with wire from a coat hanger (or use a bicycle spoke).

Place in the sun.

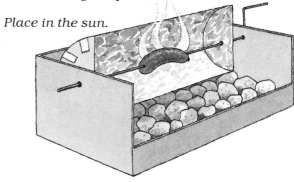

Who can make the most efficient solar cooker?

Paper making

Take six sheets of white toilet paper and six sheets of pink.

Add the sheets to about a litre of water.

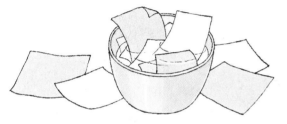

Stir thoroughly to break up all the fibres in the toilet paper.

Take a piece of fine wire gauze. Alternatively, make a tray by stretching a square from a pair of ladies' tights over a wooden frame.

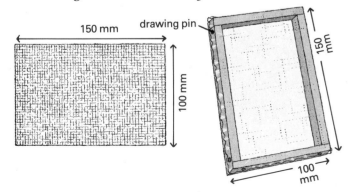

150 mm

drawing pin

100 mm

150 mm

100 mm

Dip the wire gauze into the toilet paper mixture.

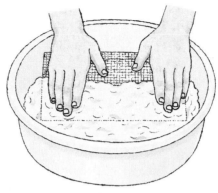

Lift the wire gauze and gently shake it so that the fibres settle and the water drips away.

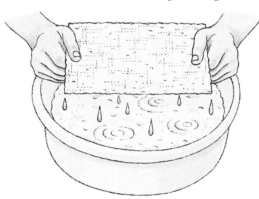

Invert the wire gauze and press the fibres into a piece of felt.

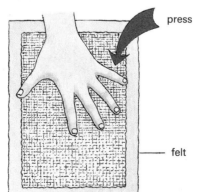

press

felt

Gently remove the gauze.

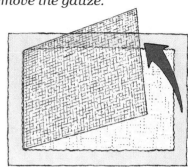

Cover the sheet of fibres with a second piece of felt. Roll firmly.

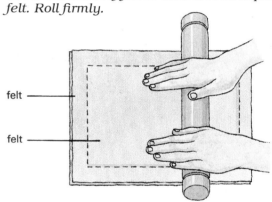

felt

felt

Gently remove the top sheet of felt.

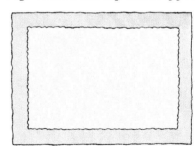

Leave to dry.

Make your own paper from:

newspapers	plant material:
blotting paper	iris leaves
magazines	rushes
paper towels	dried grasses

Different types of paper

Collect papers and make a display. Thick, thin, plain, coloured, glossy, glazed, dull, crinkly, stiff, translucent, speckled, and so on.

newspaper

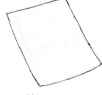 *writing paper*

 brown paper

carbon paper

 *kitchen paper*

 *tissue paper*

Test each paper.

How easy is it to write on?

Try these

pencil   Biro   crayon   felt tip

How well will each paper soak up water?

How easily does each paper tear?

Is each paper rough or smooth? Is there a paper which is smooth one side and rough on the other?

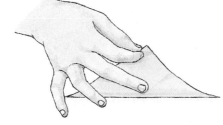

Sorting games

Can you sort paper

from smoothest to roughest?
from thickest to thinnest?
on its transparency?

Which of the daily papers is most transparent?

Papier-mâché

Papier-mâché figures are good for illustrating how the paper fibres in a flimsy structure, such as newspaper, can be transformed into a rigid form.

Mix up some non-fungicide wallpaper paste.

Add torn-up newspaper. Use small pieces.

Soak and stir.

Squeeze the water from the pulpy mass.

Make puppet heads to use in the theatre on page 44.

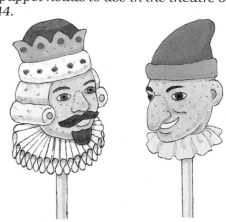

Make a collection of bricks

Are some bricks heavier than others?

How much water do bricks absorb?

Check the mass each day until it becomes constant. Use different types of brick.

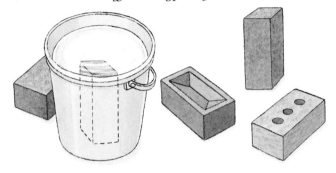

How rapidly does water rise in a brick?

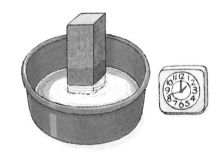

Compare the rate of rise of water in different bricks.

Rising damp

What materials can be used to provide a damp-course?

Try

> card
> wood
> slate
> paint
> varnish
> cloth
> plastic sheet

Look at the brick pattern in buildings

How many different brick patterns can children spot on their way to school?

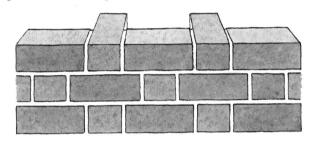

Flemish bond

English bond

Stretcher bond

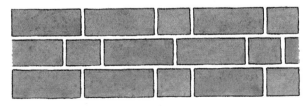

Monk bond

How strong are walls?

Are some bondings stronger than others?
Using wooden blocks, test these
   stretcher bond
   Flemish bond
   English bond
   Monk bond

Test double walls (again made from wooden blocks).

You may need to invent a strong device to smash some of your walls.

Mortar

Make different mortar mixes.

Always wear gloves.

matchbox moulds

Does the age of the mixes make a difference?
Does the amount of water used matter?

Try these ratios.

Sand	Cement
1	1
1	2
2	1
1	3
3	1
1	4
4	1

Testing mortar samples

Drop in the playground.

How many drops does it take to break each mixture?

Invent strength testing devices.

This one uses a postal tube and a 500 g mass.

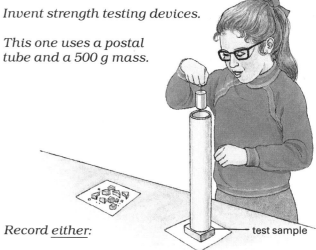

test sample

Record <u>either</u>:

how many drops it takes to break the test sample

<u>or</u>

the height from which the mass has to be dropped in order for it to break the sample.

Rubber balls

Measure the bounce from different heights.

Height dropped	Height of bounce
2·00 m	
1·75 m	
1·50 m	

Keep height constant but drop on different surfaces.

Surface	Height of bounce
stone	
concrete	
wood	
carpet	
polystyrene	
foam rubber	
grass	
PE mat	

Elastic bands

Load the tub gradually with small weights.

Plot the increase in the length of the elastic band against the increase in mass.

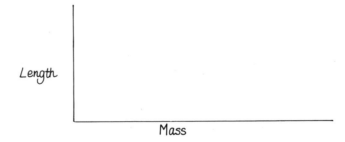

Length

Mass

Mind your toes, the elastic band may break suddenly!

Put some bands in the refrigerator. How do frozen bands fare when you test them?

Warm some elastic bands on a radiator or with a hair dryer. How do these fare?

Elastic band balance

rear view

Make your own arbitrary scale or graduate the band against a known set of weights.

Try leaving the balance permanently stretched by a weight for a week. The elastic band will not then return to its original position. This phenomenon is known as creep.

Musical bands

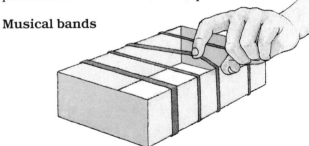

Do thin elastic bands give different notes to thick ones?

Does the length of the band make a difference to the sound?

Can you get high notes?
                  low notes?
                  soft notes?
                  loud notes?

Elastic band power

Make a simple mangonel.

Take care!

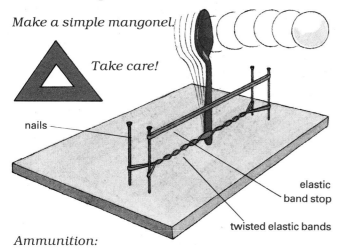

nails

elastic band stop

twisted elastic bands

Ammunition:

table-tennis ball   Gamster ball   Plasticine

Make 1 g, 2 g, 4 g and 8 g Plasticine missiles.

Mass of missile	Firing distance

Have a competition to find who can make a mangonel to fire a Plasticine missile, of given mass, the greatest distance.

Children will need to consider these variables:

thick bands or thin bands

firing release position

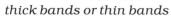

kind of spoon

number of twists in the elastic band

Springy things

Examine the springiness of wood and metal. Plot mass against amount of bending.

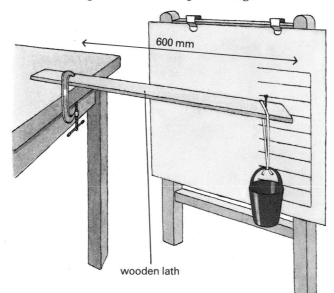

600 mm

wooden lath

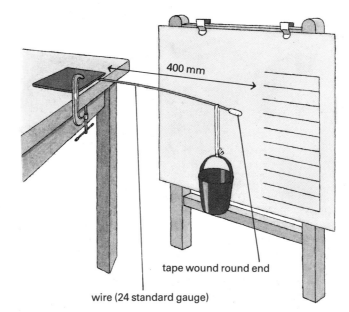

400 mm

tape wound round end

wire (24 standard gauge)

Make some spring balances.

scale (will show as plate descends)

plastic plate

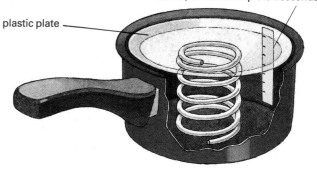

Collect engineering springs.

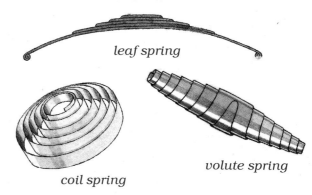

leaf spring

volute spring

coil spring

What materials would make the best surfaces for worktops and floors?

Make a collection of different materials.

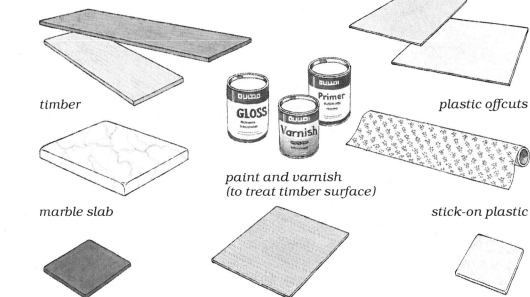

timber

marble slab

quarry tiles

paint and varnish
(to treat timber surface)

carpet tiles

plastic offcuts

stick-on plastic

ceramic tiles

Devise tests.

Spill things on different surfaces.

How do different surfaces cope with spills?

Do the spilt substances soak in?

Which is easy to mop up?
Which leaves a stain?

Can stains be easily removed?

Spill things on floors.

Try

milk
orange juice
tea
tomato sauce
vinegar

Which floors cope best with kitchen spillages?

Do cleaning liquids help?

Do they stand up to scouring?

Do hot things leave a mark?

Can surfaces withstand scratching?

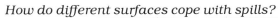
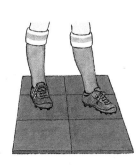

How hard wearing?

Drop a variety of things on different floor materials.

Rub with a piece of broken brick.

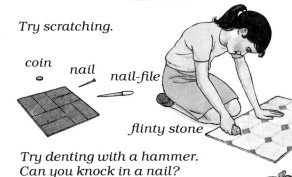

Try scratching.

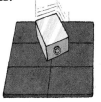

coin *nail* *nail-file*

flinty stone

Try denting with a hammer. Can you knock in a nail?

Children will need to think out how they are going to make all these tests fair.

How slippery?

Shiny floors and wet floors can be dangerous. Test floors for slipperiness.

With main body weight on one leg, slide the other leg over the floor.

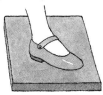

quarry tile   *rubber mat*   *concrete*

carpet tile   *lino*

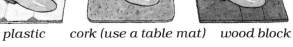

plastic   *cork (use a table mat)*   *wood block*

Do it wearing

leather shoes   *trainers*

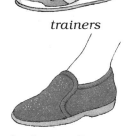

wellingtons   *bedroom slippers*

Can you put the floors in order of grip?

Measure the force.

Test shoes with soles of

> leather
> rubber
> plastic

Test well-worn soles.

Test the shoes on

> plastic
> wood
> carpet
> linoleum
> cork
> concrete
> etc.

Keep a record.

Type of sole	Type of floor	Force

Plastic bags

Collect a variety of plastic carrier bags used to take home groceries.

Which is easier on the hand?

Load each with a brick or with sand. Walk around the playground. How comfortable is each handle to hold?

Measure the 'cut' of each handle.

If the handles of the bags vary in thickness, you could use tape looped from each bag for a fair test.

Which carries the heaviest load?

Cut a strip of plastic from each bag.

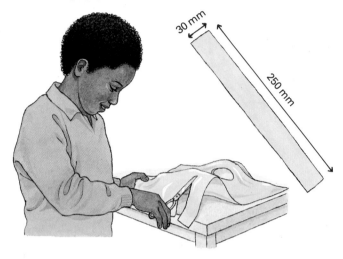

Tape one end of the strip to a dowel rod.

Tape the other end to a pencil.

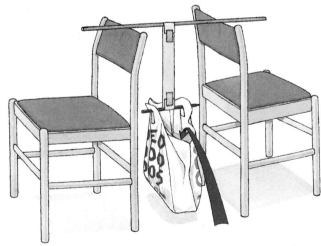

Pour sand into the plastic bag.

Measure the stretch to breaking point of the strips cut from various plastic bags.

Plastic spoons

Which is best: a plastic, metal or wooden spoon?

Test for heat conduction with hot water.

Feel the handles frequently. Which warms up the quickest?

Design a plastic bag

Make a plastic bag from sheet plastic, staples and sticky tape, to hold these items.

orange juice

bread

six apples

lemonade

Whose design holds the contents best? Is it the most attractive bag?

Plastic containers for controlling smells

Give a blindfolded person a piece of apple to eat whilst smelling a pear.

What do they say they are eating?

The ability to 'fool' the senses can be obvious. However, in the kitchen the smells of one food can taint another.

Test tainting.

Put these pairs into separate plastic containers and replace the lids. Taste after 24 and 48 hours.

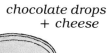

chocolate drops
+ onion

chocolate drops
+ cheese

chocolate
drops
+ mint

Are the chocolate drops tainted? Devise ways to prevent tainting.

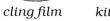

cling film   kitchen foil   plastic box

Plastic bottles for kitchen liquids

A broken squash bottle can be a nuisance, a broken bleach bottle can be harmful. Plastic bottles need to be strong.

Devise some strength tests.

Clean each bottle.
Fill each with water.
Test in a plastic bag for safety.

How many knocks?

How many turns of the vice?

How many drops?

Children will need to think about how to make each test fair.

Bottles, mugs, cups and jugs are often knocked over.

Find out how stable they are.

Begin by making a collection. Plastic ones are safest for testing.

How easily do they topple?

Make sure each vessel is toppled from the same mark.

Test for toppling by knocking

How easily are the vessels knocked over
  when empty?
  when full?

Examine tilting

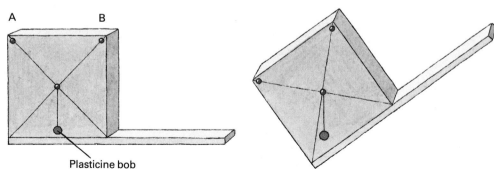

Plasticine bob

What happens to the Plasticine and string plumb line when the block is tilted?

What happens to it when it is pinned at A or at B?

Design a tea towel

Test its efficiency for wiping up

Use a

> tea towel
> hand towel
> flannel
> duster
> piece of cotton sheet
> piece of an old shirt

and so on, to dry wet dishes.

Which is best?

Test for absorbency

Put a little water into each yoghurt pot and note how quickly the different types of cloth absorb it.

cotton   wool   linen   towel   close weave   open weave

Drying time

Test a range of materials to find which dries the best. Use the 'spin dryer' to get rid of the excess water. Be careful to give each tea towel the same amount of 'spin' to make the test fair.

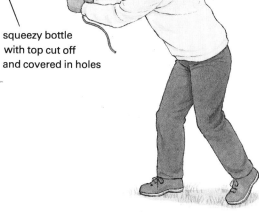

squeezy bottle with top cut off and covered in holes

When all the tests have been completed, have a competition to find who can design the best tea towel, both aesthetically and in terms of creating a good dryer.

Which washing powder washes whitest?

Test them out.
Use pieces of white cotton. Stain.

tea

tomato sauce

fruit juice

olive oil

mustard

ink

Wash the test pieces.

For a fair test children will need to think about
    amount of washing powder
    amount of water
    amount of agitation
    temperature of
    the water used
    each time.

Which is best for a white wash?

Try soaking and gently squeezing, or agitating
strongly throughout the wash.

soaking and
squeezing gently

agitating with a
wooden spoon

Try each in

cold water

hot water

cold water +
washing powder

hot water +
washing powder

Which method gives the best results?

Would this method be successful for other
materials?

Hard and soft water

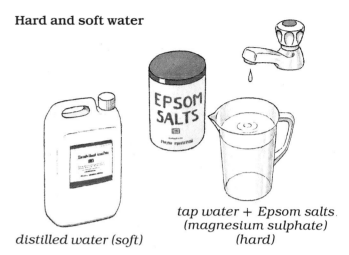

distilled water (soft)

tap water + Epsom salts
(magnesium sulphate)
(hard)

Add one soap flake at a time to each type of
water.

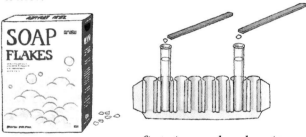

soft water   hard water

Shake each tube.

Which gives the best lather?

Invent machines to help with the washing

Here is an example of an old dolly.

Who can invent the best agitator for washing clothes in a washing bowl?

How many ways can you find to hang washing on a line?

clothes peg
(a spring and lever)

split twig
(wedge)

safety pins

tying at the corner

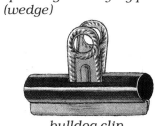

bulldog clip
(spring and lever)

Here is an old mangle.

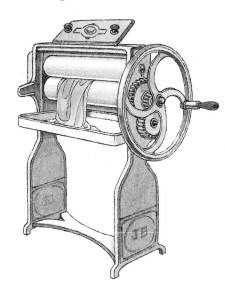

Invent other ways of getting water out of clothes

Baking bread

Make flour by grinding some corn.

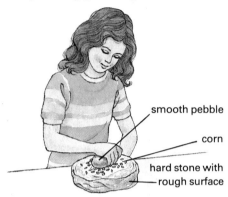

smooth pebble

corn

hard stone with rough surface

Which flours give the strongest dough?

Compare dough made from ordinary flour with that made from 'strong flour' (used for bread making).

Mix 50 g of each flour with 25 ml of water. Roll each into a ball. Make sure each ball is the <u>same size</u>.

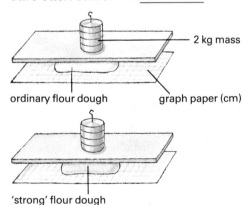

2 kg mass

ordinary flour dough

graph paper (cm)

'strong' flour dough

Count the squares over which the dough has spread at two minute intervals.

Yeast acts on the starches in the flour releasing carbon dioxide, the bubbles of which gives bread its texture.

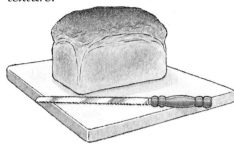

Test the action of yeast on sugar under different conditions.

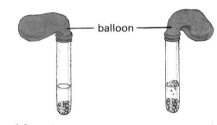

balloon

cold water
+ yeast
+ sugar

warm water
+ yeast
+ sugar

When is the most carbon dioxide produced?

Bake some bread rolls.

Recipe for bread rolls

Ingredients
200 g plain strong flour
25 g lard
1 level teaspoon salt
1 teaspoon sugar
15 g fresh yeast
125–150 ml warm water
milk

Utensils
mixing board
teaspoon
small basin
measuring jug
baking sheet
pastry brush

Method
1. Set the oven at 220 °C, Regulo 7.

2. Grease the baking sheet.

3. Put the flour in the mixing bowl.
Add the sugar, salt and lard.
Rub the lard in using your fingers.

4. Cream the yeast with two teaspoons of warm water. Add the remaining water to the creamed yeast.

5. Stir the yeast and water into the flour.

6. Using only one hand, mix the dough well until it is smooth.

7. Put the dough on a floured board.

8. Knead the dough until it is really smooth.

9. Divide the dough into eight pieces. Make them into rolls.

10. Put the rolls on to a baking sheet and leave them to rise in a warm place.

11. Once they have doubled their size, glaze the tops with milk.

12. Bake for 15–20 minutes.

13. Reduce the heat to 180 °C, or Regulo 4, after 10 minutes.

Cakes

Baking powder is a mixture of sodium bicarbonate and an acid. The acid acts on the sodium bicarbonate, releasing carbon dioxide. This rises as bubbles in the cake mixture and gives the cake its texture.

Investigate the action of acids on sodium bicarbonate.

lemon squash
(citric acid)

vinegar
(acetic acid)

water
(control)

Make some cakes using any standard recipe that requires baking powder.

Try the same recipe using sodium bicarbonate instead of baking powder, and again without any type of rising agent.

Butter

Make some butter.

Shake a half pint of double cream in a jar. Use a 'figure of eight' motion as you shake.

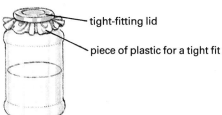

tight-fitting lid

piece of plastic for a tight fit

The fat droplets in the cream are forced together by the agitation. Larger and larger globules form until they separate from the cream, leaving a liquid called the butter fat.

Jelly

Gelatine, found in commercial jellies, is a protein. It dissolves in hot water to give a *sol*. When cool it forms a structure, beloved of children, which is called a *gel*. The gelatine forms a network of protein molecules that trap water.

Make a jelly.

Cut the jelly into equal sized cubes.

Put them in a refrigerator.

Eat one a day.

What happens as they get older?

As gelatine ages, it loses its ability to hold water. The jelly cubes in the refrigerator will become tougher as the days pass and they become drier.

Yoghurt

Make some yoghurt.

1 Heat some milk until it rises.

2 Allow to cool until luke warm.

3 Pour half the milk into a basin.

4 Add two tablespoons of plain yoghurt. Stir well.

5 Gently add the rest of the milk.

6 Pour into a jar. Wrap with a towel to keep it warm.

Leave for 8–12 hours.

Flavour with fruit.

Milk sugar is called lactose. Bacteria feed on this, producing lactic acid as a waste product. This denatures milk protein, making the milk thicker. The resultant nutty, sour tasting mixture is yoghurt.

Wordprocessing

Using the wordprocessor to construct accounts of the tests, constructions, projects, and so on, that groups of children have performed should be a collaborative exercise. Children can alter one another's writing, justify what they have written and discuss and argue as they go along.

Software

Software can be used during projects with a technological basis.

Turtle graphics

The Valiant Turtle is a common form of robot used in schools. The logical process from software to command to direct action provides a quick understanding of programs. The Turtle is controlled by the computer language called LOGO.

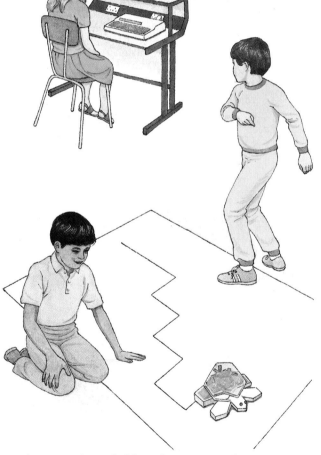

Make sure that children have enough space to move around the Turtle. They can then try walking the shape they are attempting to create, and get a feeling for how the Turtle should move.

Control technology

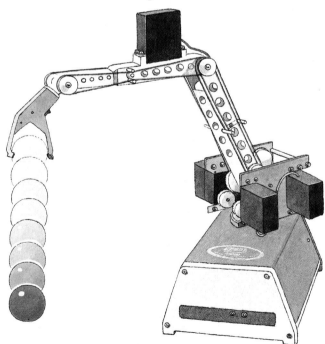

There are other robots like the Turtle that can be controlled via the computer.

Alfred Robot System

LEGO Technic Kits are excellent for making practical the principles used in technology. How levers work, the transfer of motion, gears, pulleys, rack and pinion and universal joints, and much more, can be 'understood' by children constructing models.

Buggy and manual controller

The buggy can be used to demonstrate vehicle control by the use of the manual. It can also be controlled by the computer using a suitable interface.

The manual introduces basic principles of control. It can control one or several mechanical processes at a time.

Concepts such as functions, sequence, program, and systems are made practical for children to examine and discuss.

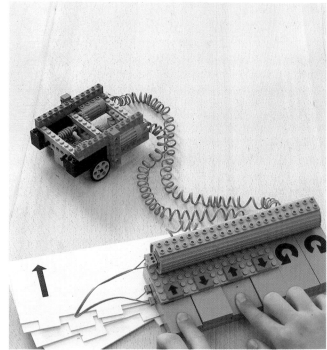

The whole dance is performed using voice and foot rhythms on a wooden floor.

Key

 strong stamp followed by three lighter stamps.

jump on to foot strongly on first beat followed by three lighter quicker steps.

Sh–mmm–Sh–mmm   voice sound by all four pistons.

Ink–a–bo–doo   voice sound for pairs.

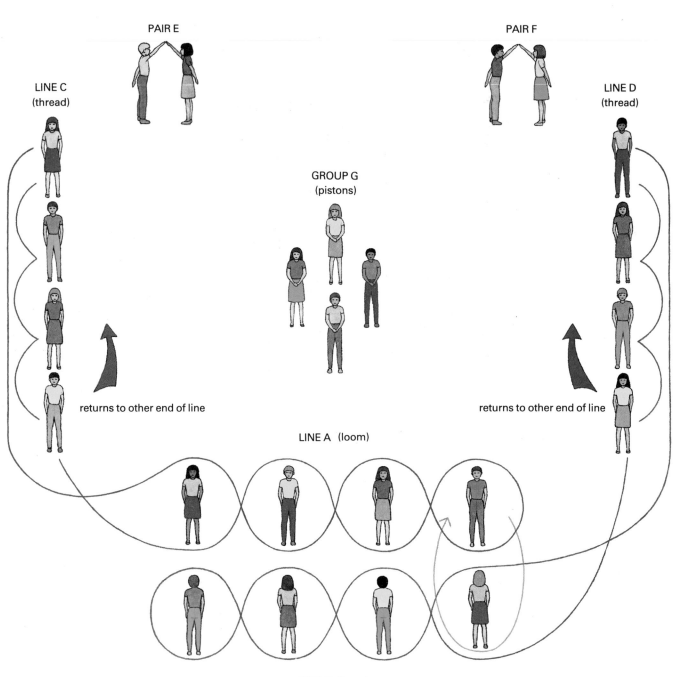

1 Lines A and B stamp ▮▮▮▮ forward to meet and back (repeat).

2 Line A stamp ▮▮▮▮ forward and behind Line B and back to place.

3 Line B stamp ▮▮▮▮ forward and behind Line A and back to place.

4 Lines A and B stamp ▮▮▮▮ forward to meet and back (repeat).

5 Line C leads line weaving through Line A. Line D leads line through Line B and all return back to original place. Foot rhythm is ▮▮▮ ▮▮▮

6 Group G clasps hands in front of bodies and push upwards and back down with rest of body in crouch near floor, whole group use voices.

7 Pair E and F move arms forward and upwards and stamp right foot forwards continually repeating movement to whole group voices. Whole dance ends with one long sh———— voice sound of stopping machines.

E. J. Arnold and Son Ltd
Dewsbury Road
Leeds, LS11 5TD

Telephone: 0532 772112

Griffin and George Ltd
Bishops Meadow Road
Loughborough
Leicestershire, LE11 0RG

Telephone: 0509 233344

Philip Harris Ltd
Lynn Lane
Shenstone
Staffordshire, WS14 0EE

Telephone: 0543 480077

Osmiroid International Ltd
Fareham Road
Gosport
Hampshire, PO13 0AL

Telephone: 0329 232345